Solution of Problems
in Control Engineering

Volume One: Linear Systems

Solution of Problems in Control Engineering

Volume One: Linear Systems

L. BALMER, B.SC.
Lecturer in Electrical Engineering, Lanchester Polytechnic

C. P. LEWIS, B.SC., C.ENG., M.I.E.E.
Lecturer in Electrical Engineering, Lanchester Polytechnic

PITMAN PUBLISHING

First published 1970

SIR ISAAC PITMAN AND SONS LTD.
Pitman House, Parker Street, Kingsway, London, W.C.2
P.O. Box 6038, Portal Street, Nairobi, Kenya

SIR ISAAC PITMAN (AUST.) PTY. LTD.
Pitman House, Bouverie Street, Carlton, Victoria 3053, Australia

PITMAN PUBLISHING COMPANY S.A. LTD.
P.O. Box 9898, Johannesburg, S. Africa

PITMAN PUBLISHING CORPORATION
6 East 43rd Street, New York, N.Y. 10017, U.S.A.

SIR ISAAC PITMAN (CANADA) LTD.
Pitman House, 381–383 Church Street, Toronto, 3, Canada

THE COPP CLARK PUBLISHING COMPANY
517 Wellington Street, Toronto, 2B, Canada

ISBN: 0 273 31411 4
(Paperback edition 0 273 31412 2)

MADE IN GREAT BRITAIN AT THE PITMAN PRESS, BATH
GO—(T.233)

Contents

Preface

It has been our intention in preparing this book not simply to give a collection of problems with answers, but by means of simple introductory examples and continuity throughout the chapters to provide an introduction to linear control system theory. Hence we have not hesitated to use problems from a wide range of examination papers or to introduce original questions where necessary.

In order to improve the understanding of the subject a few of the problems have been analysed by more than one method—we feel sure that the reader would benefit by approaching other problems in this manner.

Although the subject is treated primarily from an electrical engineer's point of view, we have used questions from other disciplines and we feel that some of the material will be of use to non-electrical engineers.

The standard of the problems is such that the needs of a wide range of students are covered, and the book should be of use in preparation for H.N.C., H.N.D., C.E.I. Part II and B.Sc. examinations.

SI units have been used throughout, and this has entailed converting some of the questions from their original form. We feel that the book has benefited from this use of a consistent system of units.

We have resisted the temptation to include questions on more advanced topics as this would have required sacrificing many of the introductory problems if the size of the book was to be kept within reasonable limits. For the same reason, introductory material to each chapter only contains the essential features of the topics under consideration, there being ample texts available to enlarge upon the subject.

To facilitate the plotting of frequency responses, a table of moduli and arguments of complex numbers has been provided as an appendix; and for use with tracing paper in solving the appropriate problems, a Nichols chart will be found in a pocket inside the back cover.

We wish to thank the many people who have helped us to obtain a wide selection of examination papers and also those who have entered into correspondence with us over points arising from these papers. In addition, we should like to thank the publishers for their many helpful comments concerning the preparation of the book.

L.B.
C.P.L.

Acknowledgments

Grateful acknowledgment is made to the following bodies for permission to reproduce questions from their past examination papers. In some cases questions have been modified by conversion to SI units; such questions are marked with an asterisk. These bodies are in no way committed to approval of the modifications or the solutions given in this book.

Brunel University
Queen's University of Belfast
University of Birmingham
University of Cambridge
University of Edinburgh
University of Leeds
University of Liverpool
University of London (Queen Mary College)
University of Manchester
University of Nottingham
University of Sheffield
University of Southampton
University of Technology, Loughborough
University of Wales:
  University College of Bangor
  University College of Swansea
  University Institute of Science and Technology, Cardiff
City of Leicester Polytechnic
Hatfield Polytechnic
Lanchester College of Technology, Coventry
Nottingham Regional College of Technology
Rugby College of Engineering Technology
Bournemouth College of Technology
Institution of Electrical Engineers
Institution of Electronic and Radio Engineers
Scottish Association for National Certificates and Diplomas

List of Principal Symbols

Conventions

A small letter represents a function of time, e.g. $f(t)$ or simple f.

A capital letter represents a function of complex frequency, e.g. $F(s)$ or simple F.

A small letter for a function with the variable included in brackets, e.g. $f(s)$, represents a function of that variable which holds only for small signals.

C	Controlled variable
E	Error $(R - C)$
G	Forward-path transfer function
H	Feedback-path transfer function
K	A constant; the gain constant in a transfer function
M	Closed-loop magnification as a function of frequency
M_m or M_{max}	Maximum value of M
M_p	Peak value of output quantity as a function of time
R	Reference variable
s	Complex frequency: $s = \sigma + j\omega$
t	Time
t_p	Time to reach the value of M_p
ζ	Damping ratio
σ	Real part of complex frequency
ω	Angular frequency; imaginary part of s; angular velocity
ω_d	Damped natural frequency
ω_m	Angular frequency at which M_m occurs
ω_n	Natural angular frequency
τ	Time-constant

Chapter 1

The Transfer Function

The *transfer function* of a system is a mathematical expression relating the output of the system to an input quantity. For a complicated system there may be several output quantities depending upon the object of the system design; for example, it may be necessary to control the angular velocity of a motor, or for an identical motor the quantity of interest may be the armature current.

Usually these quantities are related by a differential equation involving time. This can be converted into an algebraic equation in s (s being the *Laplace operator*) by use of the *Laplace transform.* However, if the initial conditions are zero an identical expression may be obtained by using s to represent a differentiation with respect to time.

Thus the general form of the transfer function is

$$\frac{\text{Laplace transform of output quantity}}{\text{Laplace transform of input quantity}} = K\frac{N(s)}{D(s)}$$

where $N(s)$ and $D(s)$ are polynomials in s, and K is a constant. In this context the Laplace operator s can be interpreted as a *complex frequency*: $s = \sigma + j\omega$.

In electrical systems the transfer function is more easily obtained if the concept of impedance is extended to include *operational impedance*. For a pure inductor the relationship between voltage and current is given by

$$v(t) = L\frac{di(t)}{dt}$$

Writing $d/dt = s$, this expression becomes

$$V(s) = Ls\,I(s)$$

where voltage and current have been written $V(s)$ and $I(s)$ to emphasize that they are here functions of s. To simplify the mathematical expressions such terms will be written V and I where there is no possibility of confusion.

The ratio $V(s)/I(s)$ is defined as the *operational impedance*, $Z(s)$; hence, for the inductor,

$$Z(s) = \frac{V(s)}{I(s)} = sL$$

Corresponding reasoning for a pure capacitor gives $Z(s) = 1/sC$, while for a pure resistor $Z(s) = R$. For sinusoidal signals the impedance $Z(j\omega)$ is obtained by the substitution $s = j\omega$ (i.e. $\sigma = 0$).

Thus the steps required in obtaining the transfer function are

1. Write down the differential equations which represent the system.
2. For ease of manipulation, replace terms involving d/dt by s, etc.
3. Eliminate all but the desired variables.

In the following problems examples of purely electrical, electro-mechanical and electro-hydraulic systems have been considered.

Example 1.1

Derive the transfer function of the circuits shown in Fig. 1.1.

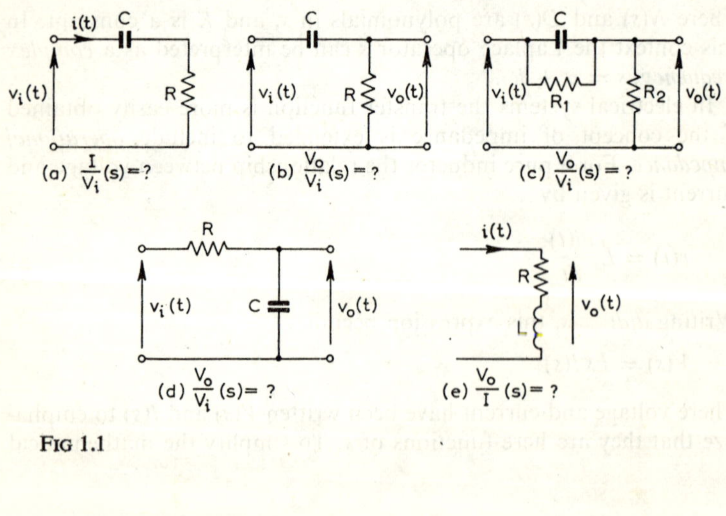

FIG 1.1

Solution

(*a*) The operational impedance of the circuit is $R + 1/sC$; hence the current I is given by

$$I = \frac{V_i}{R + 1/sC}$$

and the transfer function is

$$\frac{I}{V_i} = \frac{sC}{1 + sCR}$$

(*b*) The expression for the current is given in (*a*) and the output voltage is given by

$$V_o = RI$$

Hence the transfer function is

$$\frac{V_o}{V_i} = \frac{R}{R + 1/sC} = \frac{sCR}{1 + sCR}$$

This transfer function can be obtained by noting that the circuit behaves as a voltage divider.

(*c*) Considering this circuit as a voltage divider, we obtain

$$\frac{V_o}{V_i} = \frac{R_2}{R_2 + \dfrac{R_1(1/sC)}{R_1 + 1/sC}} = \frac{R_2}{R_2 + \dfrac{R_1}{1 + sCR_1}}$$

$$= \frac{R_2}{R_1 + R_2} \frac{(1 + sCR_1)}{\left(1 + \dfrac{sCR_1R_2}{(R_1 + R_2)}\right)}$$

$$= \frac{K(1 + s\tau_1)}{(1 + s\tau_2)}$$

where

$$\tau_1 = CR_1 \qquad \tau_2 = \frac{CR_1R_2}{(R_1 + R_2)} \qquad \text{and} \qquad K = \frac{R_2}{R_1 + R_2}$$

By remembering that $s = j\omega$ for sinusoidal inputs a useful check is to obtain the d.c. transfer function by putting $s = 0$.

(*d*)

$$\frac{V_o}{V_i} = \frac{1/sC}{R + 1/sC} = \frac{1}{1 + sCR}$$

(*e*)

$$V_o = IR + sLI = (R + sL)I$$

$$\frac{V_o}{I} = R + sL$$

Example 1.2

Determine the transfer function V_o/V_i for the ladder network shown in Fig. 1.2.

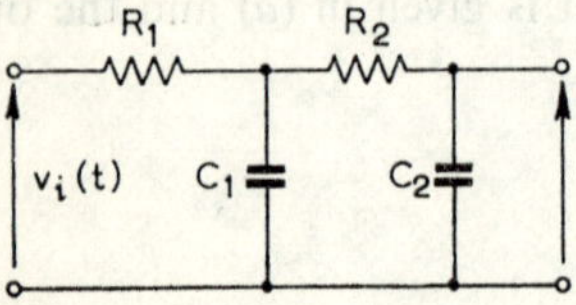

FIG 1.2

Solution
Taking the general case consider the network as shown in Fig. 1.3,

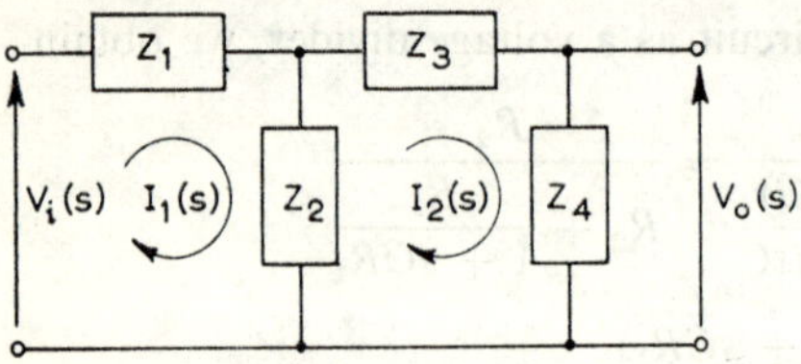

FIG 1.3

where Z_1, Z_2, etc., represent operational impedances. Applying Kirchhoff's laws,

$$V_i = (Z_1 + Z_2)I_1 - Z_2 I_2$$
$$0 = -Z_2 I_1 + (Z_2 + Z_3 + Z_4)I_2$$

Eliminating I_1,

$$\frac{V_i}{I_2} = \frac{(Z_1 + Z_2)(Z_2 + Z_3 + Z_4) - Z_2^2}{Z_2}$$

But $V_o = Z_4 I_2$, so that

$$\frac{V_o}{V_i} = \frac{Z_4 Z_2}{Z_1 Z_2 + Z_1 Z_3 + Z_1 Z_4 + Z_2 Z_3 + Z_2 Z_4}$$

Substituting $Z_1 = R_1$, $Z_3 = R_2$, $Z_2 = 1/sC_1$ and $Z_4 = 1/sC_2$, gives

$$\frac{V_o}{V_i} = \frac{1}{1 + s(R_1C_1 + R_1C_2 + R_2C_2) + s^2R_1C_1R_2C_2}$$

Example 1.3

A *damping transformer* is a device used to obtain a rate-of-change signal in electrical control systems. The equivalent circuits are shown in Fig. 1.4 for load circuits having (*a*) high input resistance, and (*b*) low input resistance.

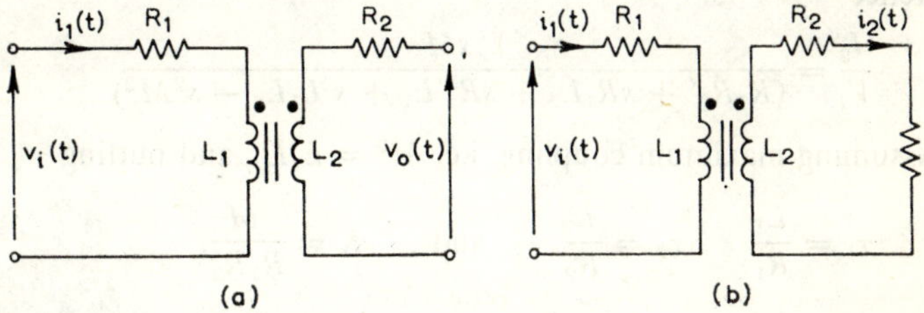

FIG 1.4

Derive the transfer functions V_o/V_i for Fig. 1.4 (*a*) and I_2/V_i for Fig. 1.4 (*b*), where R_1 and R_2 are primary and secondary winding resistances (Ω); L_1 and L_2 are primary and secondary inductances (H); and M is the mutual inductance between the windings (H).

Solution

(*a*) Considering the primary circuit,

$$V_i = (R_1 + sL_1)I_1$$

and for the secondary circuit,

$$V_o = sMI_1$$

Hence

$$\frac{V_o}{V_i} = \frac{sM}{R_1(1 + sL_1/R_1)} = \frac{Ks}{1 + s\tau}$$

where

$$K = \frac{M}{R_1} \quad \text{and} \quad \tau = \frac{L_1}{R_1}$$

(*b*) The sign of the mutually induced e.m.f. is taken as positive if the fluxes in the core of the transformer are in the same direction.

$$V_i = (R_1 + sL_1)I_1 - sMI_2$$
$$0 = (R_2 + r + sL_2)I_2 - sMI_1$$

Let the total secondary circuit resistance be R_2'; i.e.

$$R_2' = R_2 + r$$

Elimination of I_1 yields

$$V_i = \frac{(R_1 + sL_1)(R_2' + sL_2)I_2}{sM} - sMI_2$$

Hence

$$\frac{I_2}{V_i} = \frac{sM}{(R_1R_2' + sR_1L_2 + sR_2'L_1 + s^2L_1L_2 - s^2M^2)}$$

Assuming maximum coupling, i.e. $M^2 = L_1L_2$, and putting

$$\tau_1 = \frac{L_1}{R_1} \qquad \tau_2 = \frac{L_2}{R_2'} \qquad \text{and} \qquad K = \frac{M}{R_1R_2'}$$

gives

$$\frac{I_2}{V_i} = \frac{sK}{1 + s(\tau_1 + \tau_2)}$$

Example 1.4

Some measure of high-frequency compensation can be achieved in an amplifier by introducing inductance in the anode circuit as shown in Fig. 1.5. Calculate the transfer function $v_o(s)/v_i(s)$ for this circuit

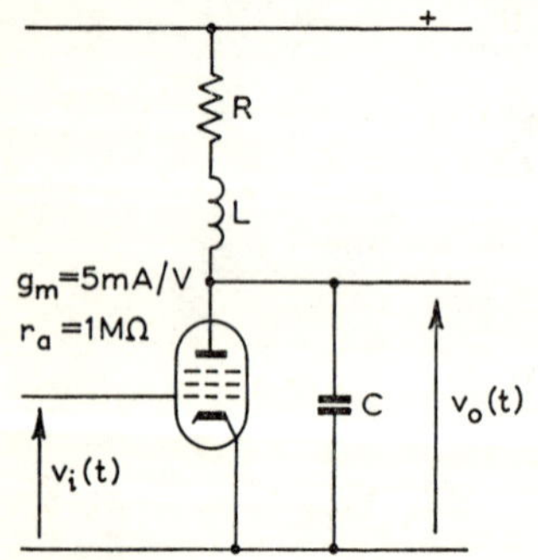

FIG 1.5

given the following component values: $R = 10\,k\Omega$, $L = 0{\cdot}5\,mH$, $C = 10pF$ (stage stray capacitance). The anode slope resistance r_a can be ignored, being much higher than the load impedance.

Solution

The equivalent circuit for the stage (ignoring r_a) is shown in Fig. 1.6. As this circuit applies to small signals only, the variables will be

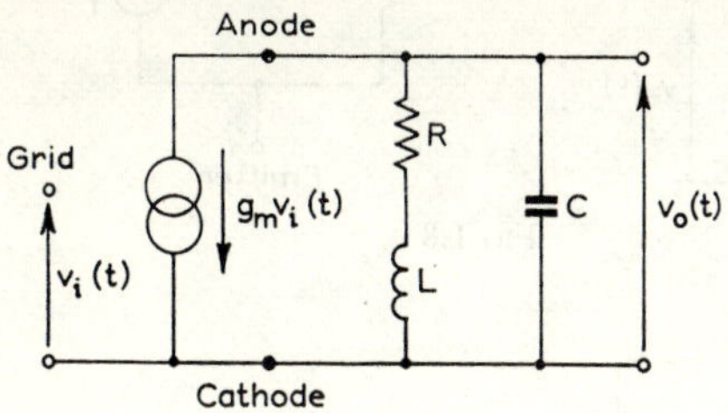

FIG 1.6

represented by small letters but shown as functions of s when they are Laplace transforms.

$$v_o(s) = -g_m v_i(s)Z(s) \tag{1}$$

where $Z(s)$ is the effective anode load operational impedance. (The minus sign occurs because of the phase reversal in the valve.)

$$Z(s) = \frac{(R + sL)(1/sC)}{R + sL + 1/sC}$$

Substitution into eqn. (1) yields

$$\frac{v_o(s)}{v_i(s)} = \frac{-g_m R(1 + sL/R)}{1 + sCR + s^2LC}$$

Inserting the numerical values given in the question,

$$\frac{v_o(s)}{v_i(s)} = \frac{-50(1 + 0{\cdot}5 \times 10^{-7}s)}{(1 + 10^{-7}s + 0{\cdot}5 \times 10^{-14}s^2)}$$

Example 1.5

A method of incorporating a lag-lead network into a transistor amplifier stage is to use the circuit shown in Fig. 1.7. Obtain the transfer function $v_o(s)/v_i(s)$ for this circuit and evaluate it, given that $R = 5\,k\Omega$, $C_1 = 5000pF$ and $R_1 = 1\,k\Omega$. (Use the simplified h equivalent circuit for the transistor taking $h_{ie} = 2\,k\Omega$ and $h_{fe} = 80$.)

Solution

The simplified h equivalent circuit for a transistor in common-emitter configuration is shown in Fig. 1.8. Here h_{ie} is the input

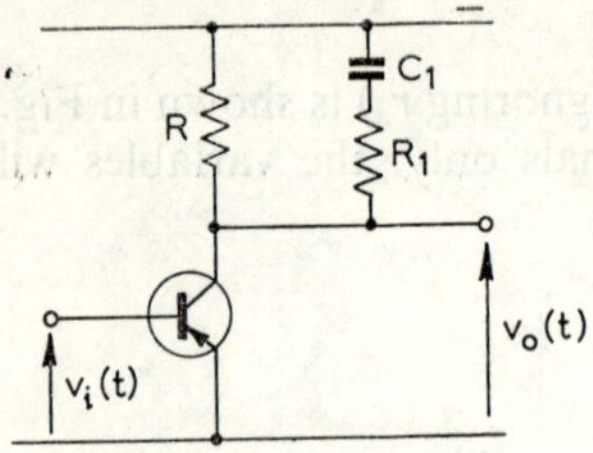

FIG 1.7

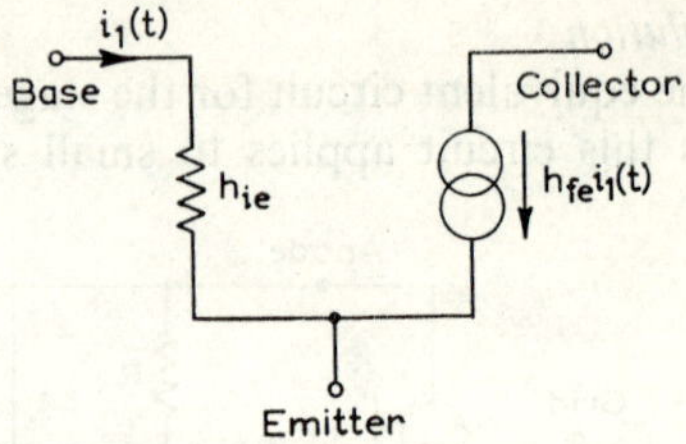

FIG 1.8

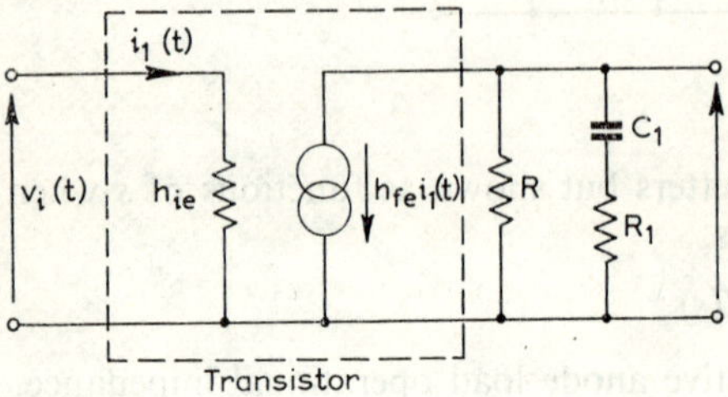

FIG 1.9

impedance with the output short-circuited to alternating current, and h_{fe} is the forward current transfer ratio with output short-circuited to alternating current. The equivalent circuit of the stage is shown in Fig. 1.9.

$$i_1(s) = \frac{v_i(s)}{h_{ie}}$$

$$v_o(s) = \frac{-h_{fe}i_1(s)R(R_1 + 1/sC_1)}{R_1 + R + 1/sC_1}$$

Therefore

$$\frac{v_o(s)}{v_i(s)} = \frac{-h_{fe}R(1 + sC_1R_1)}{[1 + sC_1(R_1 + R)]h_{ie}}$$

Inserting the values given above,

$$\frac{v_o(s)}{v_i(s)} = \frac{-200(1 + 5 \times 10^{-6}s)}{(1 + 30 \times 10^{-6}s)}$$

Example 1.6

One method of compensating for high-frequency effects in a transistor amplifier is by the correct choice of the time-constant CR_e shown in Fig. 1.10. Obtain the transfer function $v_o(s)/v_i(s)$ for this

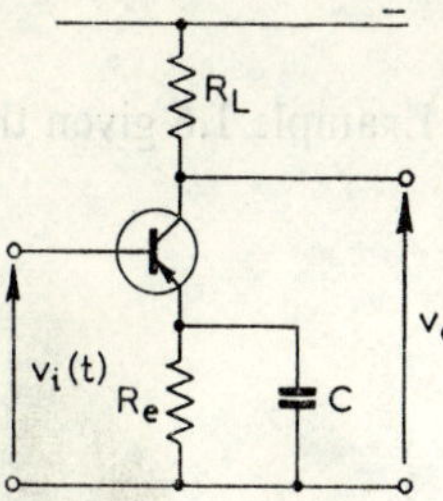

FIG 1.10

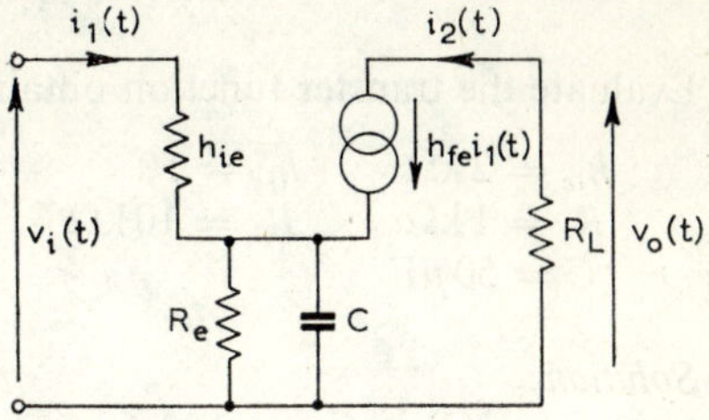

FIG 1.11

stage (use the simplified *h*-parameter equivalent circuit).

Solution
The equivalent circuit is shown in Fig. 1.11. Applying Kirchhoff's law to the input circuit,

$$v_i(s) = i_1(s)h_{ie} + [i_1(s) + h_{fe}i_1(s)]Z_e(s) \tag{1}$$

where

$$Z_e(s) = \frac{R_e}{1 + sCR_e}$$

Applying Kirchhoff's law to the output circuit,

$$v_o(s) = -i_2(s)R_L = -h_{fe}i_1(s)R_L \tag{2}$$

From eqn. (1),

$$i_1(s) = \frac{v_i(s)}{h_{ie} + (1 + h_{fe})Z_e(s)}$$

Substituting in eqn. (2),

$$v_o(s) = \frac{-h_{fe}v_i(s)R_L}{h_{ie} + \dfrac{(1 + h_{fe})R_e}{1 + sCR_e}}$$

$$\frac{v_o(s)}{v_i(s)} = \frac{-h_{fe}R_L(1 + sCR_e)}{h_{ie}(1 + sCR_e) + (1 + h_{fe})R_e}$$

$$= \frac{-h_{fe}R_L}{[h_{ie} + (1 + h_{fe})R_e]} \frac{(1 + s\tau_1)}{(1 + s\tau_2)} \tag{3}$$

where

$$\tau_1 = CR_e \quad \text{and} \quad \tau_2 = \frac{CR_e h_{ie}}{h_{ie} + (1 + h_{fe})R_e}$$

Example 1.7

Evaluate the transfer function obtained in Example 1.6 given that

$$h_{ie} = 2\,\text{k}\Omega \qquad h_{fe} = 50$$
$$R_e = 1\,\text{k}\Omega \qquad R_L = 10\,\text{k}\Omega$$
$$C = 50\,\mu\text{F}$$

Solution
Using eqn. (3) in Example 1.6, we have

$$\frac{v_o(s)}{v_i(s)} = \frac{-h_{fe}R_L}{[h_{ie} + (1 + h_{fe})R_e]} \frac{(1 + s\tau_1)}{(1 + s\tau_2)}$$

$$\tau_1 = CR_e = 50 \times 10^{-6} \times 1 \times 10^3 = 50 \times 10^{-3}\,\text{s}$$

$$\tau_2 = \frac{CR_e h_{ie}}{h_{ie} + (1 + h_{fe})R_e}$$

$$= \frac{50 \times 10^{-6} \times 1 \times 10^3 \times 2 \times 10^3}{2 \times 10^3 + (1 + 50) \times 1 \times 10^3} = 1{\cdot}89 \times 10^{-3}\,\text{s}$$

Over the frequency range concerned the h-parameters are constant and the term

$$\frac{h_{fe}R_L}{h_{ie} + (1 + h_{fe})R_e}$$

has the numerical value 9·44. Hence

$$\frac{v_o(s)}{v_i(s)} = \frac{-9{\cdot}44(1 + 50 \times 10^{-3}s)}{(1 + 1{\cdot}89 \times 10^{-3}s)}$$

Example 1.8

Derive the transfer function of a computing amplifier with computing elements as shown in Fig. 1.12.

Solution
A computing amplifier with computing elements Z_1 and Z_2 is shown in Fig. 1.13. The amplifier must operate in its linear range, and as

its gain is very large this means that the voltage across its input terminals must be small compared with V_i and V_o. Calculations are

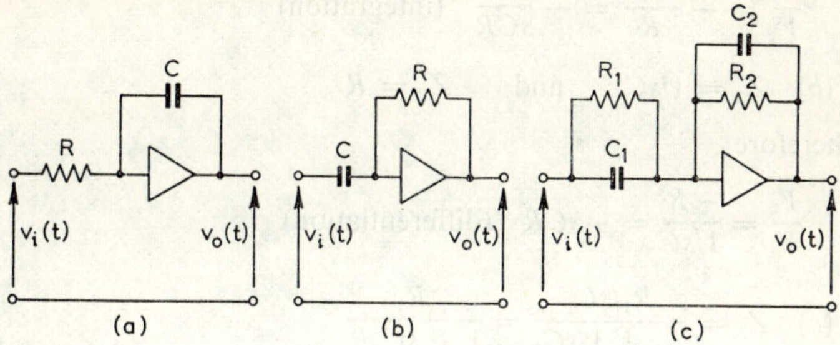

FIG 1.12

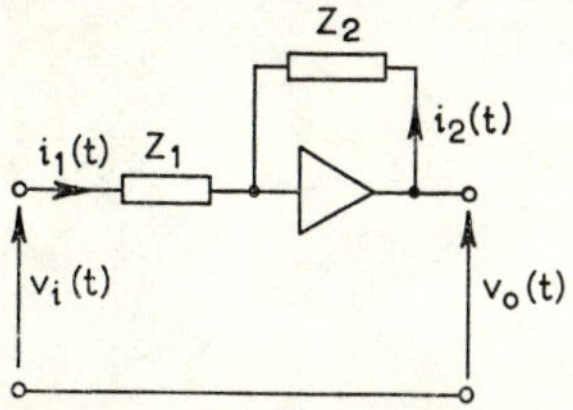

FIG 1.13

simplified if this voltage is assumed to be zero—this assumption is known as the *virtual earth principle*. It may also be assumed that the input current is negligible. Hence

$$I_1 = \frac{V_i}{Z_1}$$

and

$$I_2 = \frac{V_o}{Z_2}$$

Also $I_1 + I_2 = 0$ (input current negligible), so that

$$\frac{V_i}{Z_1} + \frac{V_o}{Z_2} = 0$$

and

$$\frac{V_o}{V_i} = -\frac{Z_2}{Z_1}$$

Applying this formula to the arrangements of Fig. 1.12:

(*a*) $Z_1 = R$ and $Z_2 = 1/sC$

Therefore

$$\frac{V_o}{V_i} = -\frac{1/sC}{R} = -\frac{1}{sCR} \quad \text{(integration)}$$

(*b*) $Z_1 = 1/sC$ and $Z_2 = R$

Therefore

$$\frac{V_o}{V_i} = \frac{-R}{1/sC} = -sCR \quad \text{(differentiation)}$$

(*c*) $$Z_1 = \frac{R_1/sC_1}{R_1 + 1/sC_1} = \frac{R_1}{1 + sC_1R_1}$$

$$Z_2 = \frac{R_2/sC_2}{R_2 + 1/sC_2} = \frac{R_2}{1 + sC_2R_2}$$

Therefore

$$\frac{V_o}{V_i} = \frac{-R_2}{R_1}\frac{(1 + sC_1R_1)}{(1 + sC_2R_2)}$$

Example 1.9

A separately excited d.c. generator has the following open-circuit characteristic:

E.M.F. (V)	7·6	120	235	340	405	445	465	477
Field current (A)	0	5	10	15	20	25	30	35

The field circuit has resistance 2·9 Ω and inductance 7·2 H. Derive the transfer function relating open-circuit armature voltage (e.m.f.) to field-circuit terminal voltage for the linear portion of the characteristic.

Solution
A circuit diagram of the generator is shown in Fig. 1.14, and the open-circuit characteristic, in Fig. 1.15. Consider the field-winding circuit,

$$V_f = R_fI_f + sL_fI_f$$

The generator e.m.f. is given by

$$E = KI_f$$

where K is the slope of the linear portion of the open-circuit characteristic. Removing I_f from these equations,

$$\frac{E}{V_f} = \frac{K}{R_f(1 + s\tau_f)}$$

where $\tau_f = L_f/R_f$, the field-circuit time-constant. Here $R_f = 2{\cdot}9\,\Omega$,

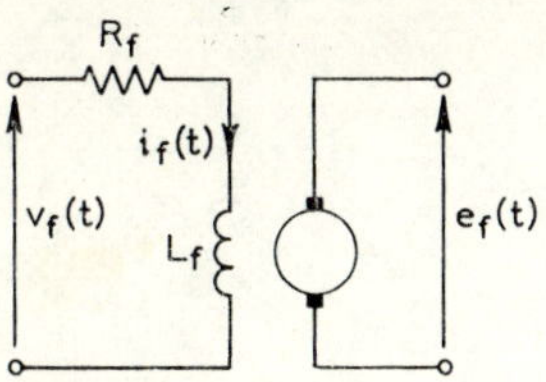

FIG 1.14

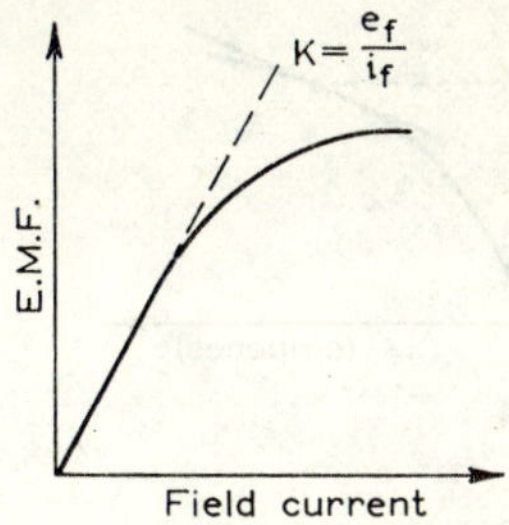

FIG 1.15 Open-circuit characteristic

$L_f = 7{\cdot}2\,\text{H}$ and from the graph the slope of the linear portion of the characteristic is 22 V/A. Substituting these numerical values,

$$\frac{E}{V_f} = \frac{7{\cdot}7}{1 + 2{\cdot}5s}$$

Example 1.10

If the field-winding inductance of the machine of Example 1.9 is assumed to be directly proportional to the slope of the open-circuit characteristic, derive the transfer function for a small variation of e.m.f. about 450 V.

Solution
The open-circuit characteristic has been redrawn in Fig. 1.16. A tangent to the curve at 450 V has the slope 6·6 V/A.

$$\frac{\text{New inductance}}{\text{Original inductance}} = \frac{\text{New slope}}{\text{Original slope}}$$

Therefore

$$\text{New inductance} = \frac{7{\cdot}2 \times 6{\cdot}6}{22{\cdot}4} = 2{\cdot}12\ \text{H}$$

and the new time-constant is 0·73 s.

Thus the transfer function for small signals when the generator is working at 450 V is

$$\frac{e(s)}{v_f(s)} = \frac{2.3}{1 + 0{\cdot}73s}$$

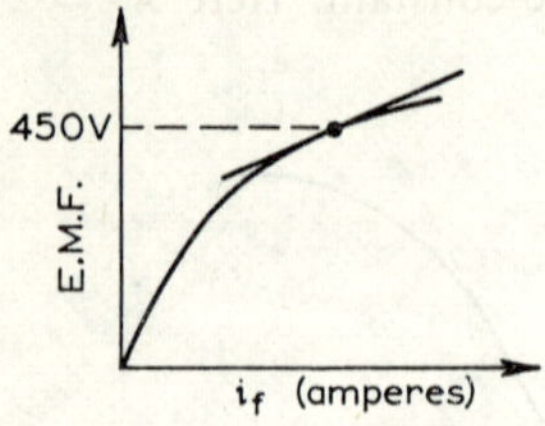

FIG 1.16

Example 1.11

The *amplidyne* is a rotary amplifier. It can be considered to be two cascaded d.c. generators—an equivalent circuit is shown in Fig. 1.17.

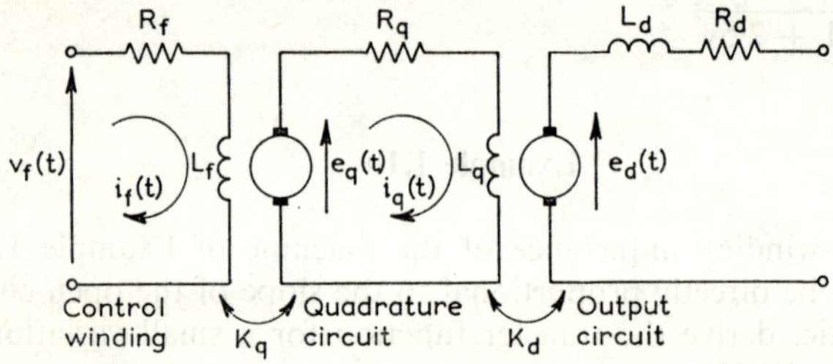

FIG 1.17

Determine the relation between the voltage V_f applied to the control-field winding and the output e.m.f. E_d.

Solution

Consider the control-winding circuit:

$$V_f = R_f I_f + sL_f I_f \quad (1)$$

For the quadrature circuit the total inductance and resistance are represented by L_q and R_q respectively. The generated e.m.f. E_q is proportional to the control-field current I_f. Hence

$$E_q = K_q I_f = R_q I_q + sL_q I_q \quad (2)$$

Similarly for the output circuit the e.m.f. E_d is proportional to the quadrature-circuit current:

$$E_d = K_d I_q \tag{3}$$

Combining eqns. (1), (2) and (3),

$$E_d = \frac{K_d K_q V_f}{(R_q + sL_q)(R_f + sL_f)}$$

The transfer function is

$$\frac{E_d}{V_f} = \frac{K_d K_q}{R_q R_f(1 + s\tau_f)(1 + s\tau_q)}$$

where

$$\tau_f = L_f/R_f \qquad \text{and} \qquad \tau_q = L_q/R_q$$

Example 1.12

A printed-armature d.c. servo motor has the torque/speed characteristic shown in Fig. 1.18. The moment of inertia of the rotor is

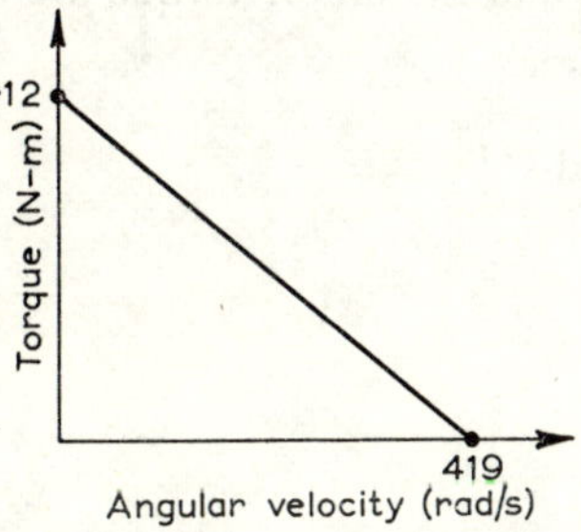

FIG 1.18

1·28 × 10^{-4} kg-m². Use this information to obtain the mechanical time-constant of the motor.

Solution

The torque/speed characteristic can be considered to be due to a motor delivering a constant torque connected to a viscous damper, i.e. a device requiring a torque increasing with speed. Thus the overall characteristic can be represented as in Fig. 1.19. The equation of motion is

$$T - B\omega = Js\omega$$

where ω is the angular velocity of the motor and B is the damping torque constant.

The transfer function from torque to angular velocity is

$$\frac{\omega}{T} = \frac{1}{B(1 + s\tau_o)}$$

where $\tau_o = J/B$ is the mechanical time-constant. For the characteristic shown in Fig. 1.18 the value of the damping constant is given

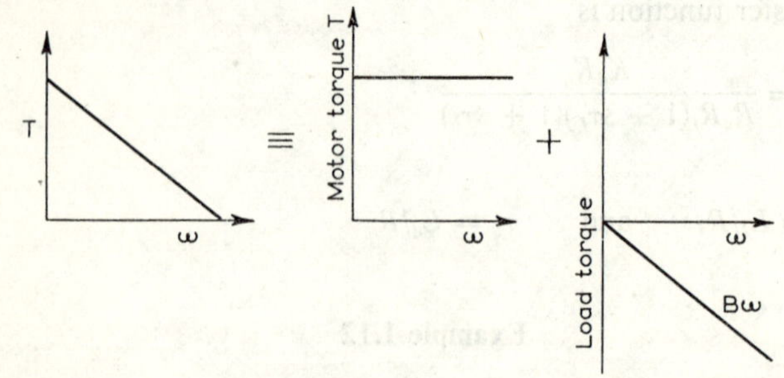

FIG 1.19

by the fact that the viscous torque and the motor torque are equal and opposite at 419 rad/s. Hence

$$B = \frac{2{\cdot}12}{419} = 5 \times 10^{-3}\,\text{N-m/rad-s}^{-1}$$

The mechanical time constant is

$$\frac{1{\cdot}28 \times 10^{-4}}{5 \times 10^{-3}} = 25\,\text{ms}$$

Example 1.13

Explain the term *transfer function* as applied to an element of an electrical control system.

A separately excited unloaded d.c. motor has armature resistance and inductance R_a and L_a, field resistance and inductance R_f and L_f, and J is the moment of inertia of its rotor. Develop expressions for the transfer function between the angular position of the shaft and (i) the applied armature voltage assuming a constant field current, and (ii) the field voltage assuming a constant armature current. Air-gap flux in the motor may be assumed proportional to the field current. (*I.E.E.*)

Solution

The motor with its associated parameters is shown in Fig. 1.20.

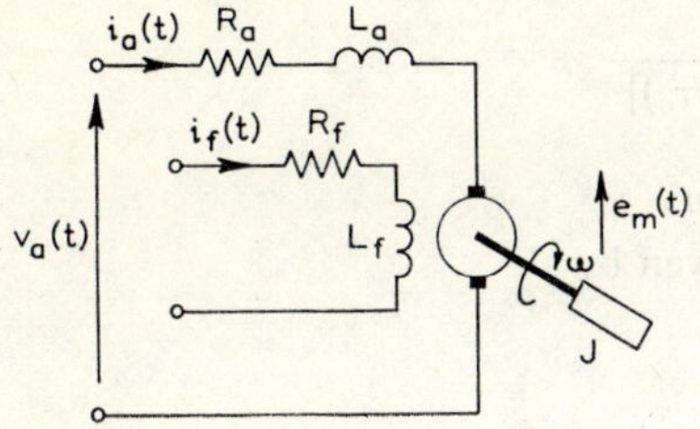

FIG 1.20

(i) *Constant field current*

Let K_V be the constant relation between e.m.f. and speed of the motor (V/rad-s^{-1}), and K_T the constant relation between motor torque and armature current (N-m/A). (Note that in SI units $K_V = K_T$ for the ideal situation.)

The electrical relations are

$$V_a = I_a R_a + sL_a I_a = I_a R_a(1 + s\tau_a) \tag{1}$$

where $\tau_a = L_a/R_a$ denotes the electrical time-constant of the armature circuit.

The motor torque and e.m.f. equations are

$$T = K_T I_a \tag{2}$$

$$E_m = K_V \omega \tag{3}$$

The equation of motion is

$$T = Js\omega \tag{4}$$

where J has the units kg-m^2. Hence from eqns. (2) and (4),

$$K_T I_a = Js\omega$$

Substituting for I_a from eqn. (1),

$$\frac{\omega}{V_a} = \frac{1}{K_V[1 + s\tau_0(1 + s\tau_a)]}$$

where

$$\tau_0 = \frac{JR_a}{K_T K_V}$$

is the mechanical time-constant of the system.

The required transfer function is between shaft position Θ and applied voltage V_a. Using the relation $\Theta = \dfrac{\omega}{s}$,

$$\frac{\Theta}{V_a} = \frac{1}{K_V s[1 + s\tau_0(1 + s\tau_a)]}$$

(ii) *Constant armature current*

The field circuit current is given by

$$I_f = \frac{V_f}{R_f(1 + s\tau_f)}$$

where $\tau_f = L_f/R_f$.

The torque of a d.c. motor is proportional to the product of the flux per pole Φ and the armature current I_a. Here $T \propto \Phi$, and we assume that $\Phi \propto I_f$, i.e. the magnetization characteristic is linear. Thus

$T = KI_f$ newton-metres per ampere

The equation of motion is

$$T = Js\omega$$

Substituting for I_f,

$$\frac{\omega}{V_f} = \frac{K}{JR_f s(1 + s\tau_f)}$$

The required transfer function is

$$\frac{\Theta}{V_f} = \frac{K}{JR_f s^2(1 + s\tau_f)}$$

Example 1.14

Derive the transfer function (ω/E_g) for the Ward Leonard loop shown in Fig. 1.21 used in a speed control system having the following specification:

Generator: 450 V, 1 000 A, 105 rad/s. Armature resistance, 0·012 Ω. Armature inductance, 1·3 mH.

Motor: 450 V, 1 000 A, 78·5 rad/s. Armature resistance, 0·018 Ω. Armature inductance, 1·8 mH. Total inertia at motor shaft, 254 kg-m^2.

Solution

The loop inductance is 3·1 mH, and the total loop resistance, 0·03 Ω. Hence the electrical time-constant is 0·1 s. For the motor,

$$K_V = \frac{450 - 1000 \times 0{\cdot}018}{78{\cdot}5} = 5{\cdot}5\,\text{V/rad-s}^{-1}$$

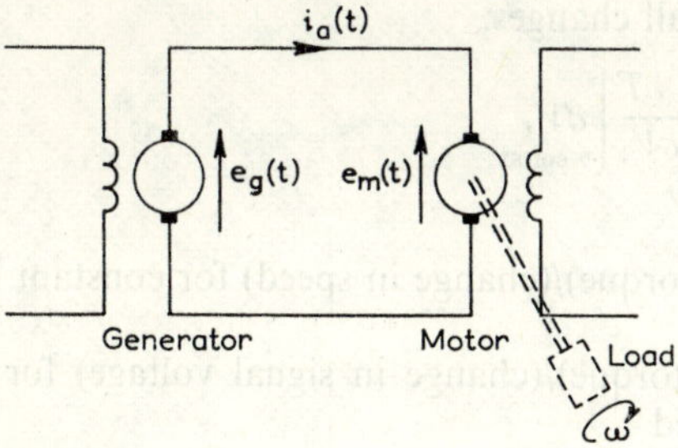

FIG 1.21

Hence the mechanical time-constant is

$$\frac{254 \times 0{\cdot}03}{5{\cdot}5^2} = 0{\cdot}25\,\text{s}$$

The transfer function is

$$\frac{\omega}{E_g} = \frac{1}{5{\cdot}5[1 + 0{\cdot}25s(1 + 0{\cdot}1s)]}$$

Example 1.15

The equivalent circuit of a 2-phase a.c. servomotor having inertia load and viscous damping is shown in Fig. 1.22 (*a*). The two windings

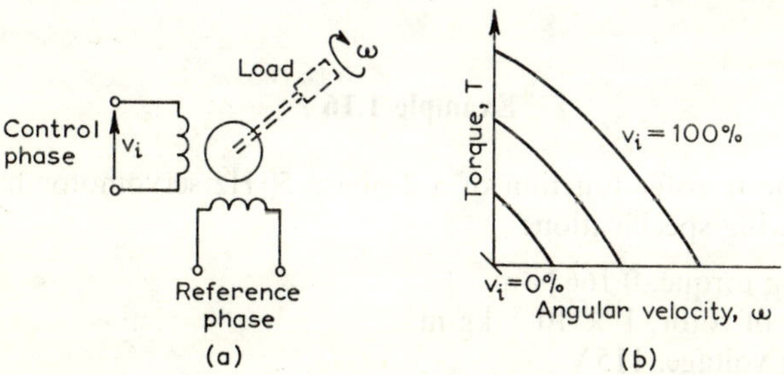

FIG 1.22

are fed from voltage sources which are in quadrature. The signal voltage supplies the control winding and the reference winding is connected to a constant-voltage source. Using the torque/speed characteristics shown, obtain the relation between the position of the motor shaft and the signal voltage.

Solution

From Fig. 1.22 (*b*), torque is a function of motor speed and signal voltage. Considering small changes,

$$dT = \left.\frac{\partial T}{\partial \omega}\right|_{V_i \text{ const.}} d\omega + \left.\frac{\partial T}{\partial V_i}\right|_{\omega \text{ const.}} dV_i$$

$$= K_1 d\omega + K_2 dV_i$$

where K_1 = (change in torque)/(change in speed) for constant input voltage

K_2 = (change in torque)/(change in signal voltage) for constant speed

Using the Laplace transform, the equation of motion is

$$Js^2\,\theta(s) + Bs\,\theta(s) = K_1 s\,\theta(s) + K_2\,v_i(s)$$

where J and B having the usual meaning and the small letters θ and v_i are used to emphasize that small signals are being considered. Hence

$$\frac{\theta(s)}{v_i(s)} = \frac{K_2}{Js^2 + (B - K_1)s}$$

$$= \frac{K'}{s(1 + s\tau_m)}$$

where

$$K' = \frac{K_2}{B - K_1} \quad \text{and} \quad \tau_m = \frac{J}{B - K_1}$$

Example 1.16

Derive the transfer function of a 2-phase 50 Hz servomotor having the following specification:

Starting torque, 0·166 N-m
Inertia of rotor, 1×10^{-5} kg-m^2
Supply voltage, 115 V
No-load angular velocity, 304 rad/s

Assume straight-line characteristics, and that the viscous friction is zero.

Solution
The torque/speed characteristic is shown in Fig. 1.23. As there is no viscous friction in this case,

$$K' = \frac{K_2}{-K_1} \quad \text{and} \quad \tau_m = \frac{J}{-K_1}$$

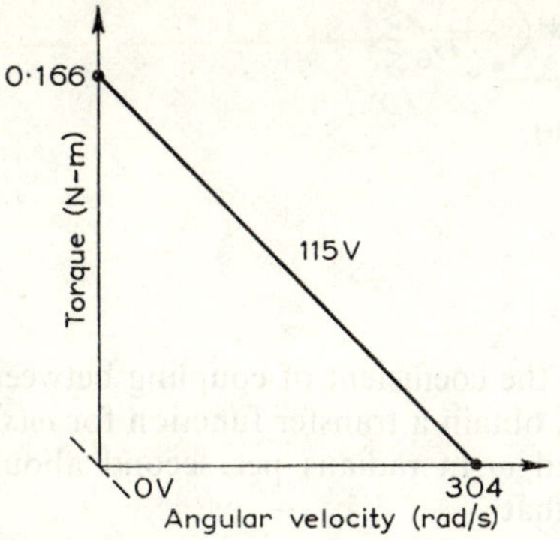

FIG 1.23

Using the information from Example 1.15,

$$K_2 = \frac{+0{\cdot}166}{115} \text{ N-m/V}$$

$$K_1 = \frac{-0{\cdot}166}{304} \text{ N-m/rad-s}^{-1}$$

Hence

$$K' = \frac{304 \times 0{\cdot}166}{115 \times 0{\cdot}166} = 2{\cdot}64$$

$$\tau_m = \frac{1 \times 10^{-5} \times 304}{0{\cdot}166} = 0{\cdot}018 \text{ s}$$

and

$$\frac{\theta(s)}{v_i(s)} = \frac{2{\cdot}64}{s(1 + 0{\cdot}018s)}$$

Example 1.17

When a split-field series d.c. motor is tested by passing a direct current of 100 mA through one field winding and the armature, the torque developed is 5×10^{-3} N-m.

The loop equation is

$$e(s) = Ri(s) + sLi(s) - sMi(s)$$
$$= [R + s(L - M)]i(s)$$

The mutual inductance of the windings is

$$M = 0{\cdot}9\sqrt{10^2} = 9\,\text{H}$$

In terms of the armature current $i_a(s)$, in milliamperes, the loop equation becomes

$$e(s) = \frac{(200 + s)i_a(s)}{2 \times 10^3}$$

As there is no change in flux the torque of the motor is directly proportional to the armature current:

$$T = Ki_a(s)$$

To obtain K we note that the machine gives a torque of 5×10^{-3} N-m when 100 mA passes through one half of the field winding and through the armature. For the condition where 100 mA passes through the whole winding and the same current passes through the armature, the torque will be double the above value. Therefore

$$10 \times 10^{-3} = 100K \qquad \text{and} \qquad K = 10^{-4}\,\text{N-m/mA}$$

Hence $T = 10^{-4}i_a(s)$.

The equation of motion is

$$10^{-4}i_a(s) = 10^{-4}s\omega(s) + 10^{-3}\omega(s)$$
$$\frac{\omega(s)}{i_a(s)} = \frac{1}{(s + 10)}\ \text{rad-s}^{-1}/\text{mA}$$

The relation between ω and e is

$$\frac{\omega(s)}{i_a(s)} \times \frac{i_a(s)}{e(s)} = \frac{1}{(s + 10)}\frac{2 \times 10^{-3}}{(s + 200)}$$

If we require output position we write

$$\theta(s) = \frac{\omega(s)}{s}$$
$$\frac{\theta(s)}{e(s)} = \frac{2 \times 10^3}{s(s + 10)(s + 200)}$$

But $e(s) = 100e_i(s)$, so that

$$\frac{\theta(s)}{e_i(s)} = \frac{2 \times 10^5}{s(s + 10)(s + 200)}$$

Example 1.18

The motor characteristics for a split field d.c. servomotor are shown in Fig. 1.26. Determine the mechanical time-constant for the values

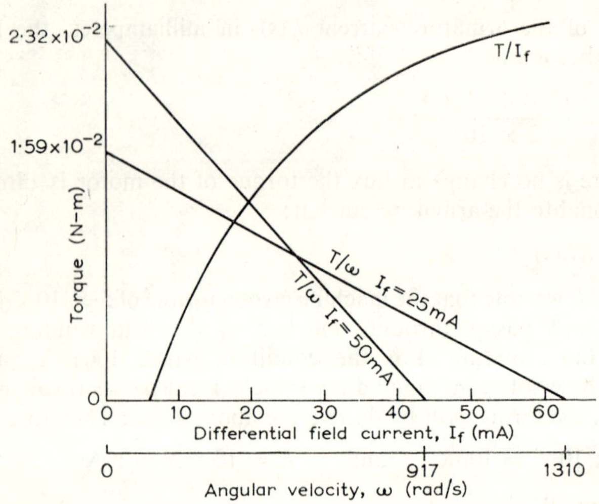

FIG 1.26

of differential field current shown (25 mA and 50mA). The moment of inertia of the armature is $1{\cdot}83 \times 10^{-6}$ kg-m². Obtain the expression for the transfer function from differential field current to motor speed at both field currents. The points on the T/I_f characteristics are as follows:

Torque, N-m $\times 10^{-2}$	0	0·49	0·98	1·47	1·96	2·45
Differential field current, mA	6	10	15·5	23	34·5	58

Solution

The values of the mechanical time-constants are obtained from the T/ω characteristics. As in Example 1.12, the value of the damping constant is given by the fact that on no-load the viscous torque must be equal to the motor torque. For 25 mA differential field current,

$$B = \frac{1{\cdot}59 \times 10^{-2}}{1{\cdot}31 \times 10^{3}} = 1{\cdot}22 \times 10^{-5}\ \text{N-m/rad-s}^{-1}$$

The mechanical time-constant is

$$\tau_0 = \frac{J}{B} = \frac{1{\cdot}83 \times 10^{-6}}{1{\cdot}22 \times 10^{-5}} = 0{\cdot}15\,\text{s}$$

K_{Tf} is obtained from the slope of the T/I_f curve at 25 mA differential field current:

$$K_{Tf} = 0{\cdot}54\,\text{N-m/A}$$

The transfer function from differential current to angular velocity is given by

$$\frac{\omega}{I_f} = \frac{K_T}{B(1 + s\tau_0)} = \frac{44{\cdot}5 \times 10^3}{1 + 0{\cdot}15s}\ \text{rad-s}^{-1}/\text{A}$$

Similarly at 50 mA differential field current,

$$\tau_0 = 0{\cdot}07\,\text{s} \qquad \text{and} \qquad \frac{\omega}{I_f} = \frac{80{\cdot}5 \times 10^3}{1 + 0{\cdot}07s}\ \text{rad-s}^{-1}/\text{A}$$

Example 1.19

A hydraulic ram often provides the power amplification in a servo system. A simplified approach is to take the flow of oil through the inlet valve to be proportional to valve opening. Neglecting the inertia of the piston, oil compressibility and leakage effects, find the transfer function between piston displacement and valve opening.

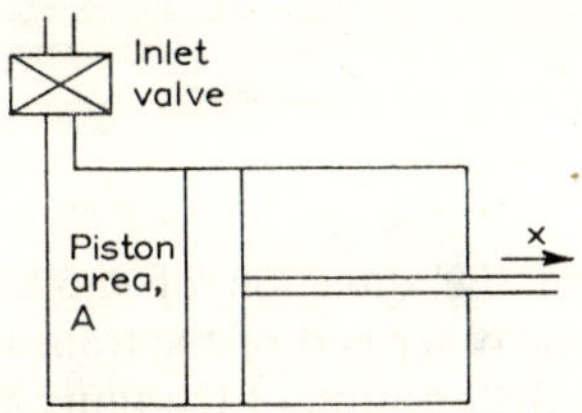

FIG 1.27

What is the transfer function for a ram having a piston area of 60 cm² and a valve giving a rate of flow of 20 cm³/s for each millimetre of valve opening?

Solution

The system is shown in Fig. 1.27. The volume of oil entering the cylinder for a displacement X is equal to the volume swept by the piston, XA.

$$\text{Flow of oil} = sXA = KY$$

where Y is the valve displacement, and K is the valve constant or flow per unit valve opening. Hence

$$sXA = KY \qquad \text{or} \qquad \frac{X}{Y} = \frac{K}{sA}$$

Inserting numerical values,

$$\frac{X}{Y} = \frac{20}{s60} = \frac{1}{3s}\ \text{cm/mm}$$

Example 1.20

Electrical feedback is often used around a hydraulic ram in order to linearize the system and reduce the order of the transfer function. A simplified block diagram of such a system is shown in Fig. 1.28.

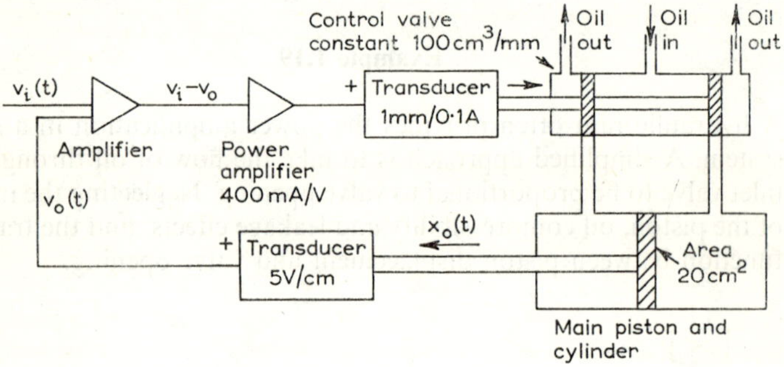

FIG 1.28

The arrows on the transducers give the direction of movement associated with a positive signal at the other end of the transducer. Obtain the transfer function relating the position of the ram, X_o, to the input voltage V_i.

Solution

If the piston is displaced a distance X_o, the volume of oil entering the cylinder will be $20X_o$ cubic centimetres.

Flow of oil $= s20X_o$

$= 100$ multiplied by control valve opening in millimetres

$$\text{Control valve opening} = \frac{(V_i - V_o)400}{100}\ \text{millimetres}$$

Therefore

$$s20X_o = 100(V_i - V_o)4$$

However, the output transducer relates V_o to X_o; i.e. $V_o = 5X_o$. Thus

$$s20X_o = 400(V_i - 5X_o)$$

and

$$\frac{X_o}{V_i} = \frac{0{\cdot}2}{1 + 0{\cdot}01s}$$

PROBLEMS

1.1 Obtain the transfer function V_o/V_i for the lead-lag network shown in Fig. 1.29. Evaluate this transfer function when $C_1 = 1\,\mu F$,

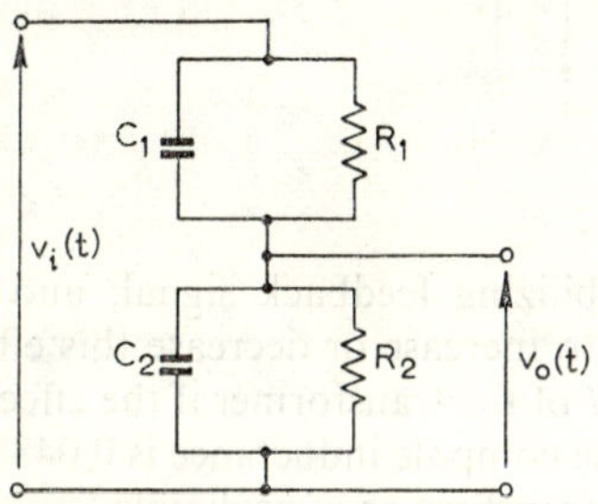

FIG 1.29

$C_2 = 0{\cdot}5\,\mu F$, $R_1 = 100\,k\Omega$, $R_2 = 10\,k\Omega$. What would be the transfer function if $C_1R_1 = C_2R_2$, assuming that the values of the resistors are unaltered?

Ans. $\dfrac{1}{11}\dfrac{(1 + 0{\cdot}1s)}{(1 + 0{\cdot}014s)}, \dfrac{1}{11}$

1.2 Obtain the transfer function V_o/V_i for the *RLC* filter shown in Fig. 1.30, where $R = 500\,\Omega$, $L = 8\,H$, $C = 50\,\mu F$.

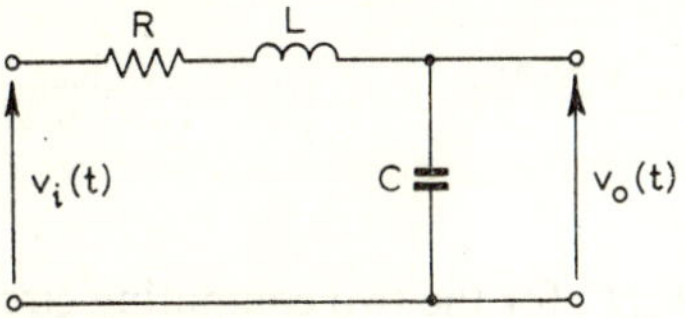

FIG 1.30

Ans. $\dfrac{1}{1 + 25s \times 10^{-3} + 4s^2 \times 10^{-4}}$

1.3 Fig. 1.31 shows a circuit used to obtain a signal proportional to a d.c. motor current which also has a rate-of-change component. The equivalent circuit is also shown. The compole resistance and inductance are R_c and L_c. A voltage proportional to the rate of

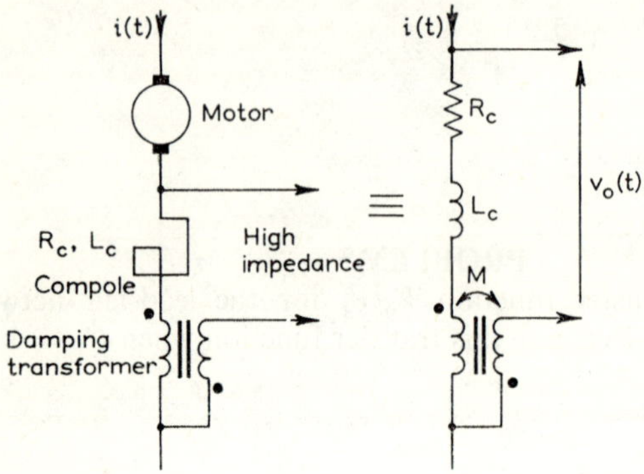

FIG 1.31

change of current is a useful stabilizing feedback signal, and the damping transformer can be used to increase or decrease this effect.

Specify the mutual inductance M of the transformer if the effective inductance required is 0·21 mH. The compole inductance is 0·048 mH and the primary resistance of the transformer is negligible.

If the primary inductance of the transformer is 0·5 μH, determine the value of the secondary inductance. Obtain an expression for $V_o(s)$ as a function of the motor current $I(s)$, assuming that the resistance of the compole winding is 0·003 Ω.

Ans. 0·16 mH, 51 mH, 0·003(1 + 0·07s)

1.4 The T equivalent circuit of a transistor is as shown in Fig. 1.32. Use this circuit to evaluate the transfer function of the stage shown in Example 1.5. Take the value of the parameters in the equivalent circuit to be as follows: $r_b = 500\,\Omega$, $r_e = 30\,\Omega$, $r_c = 1{\cdot}5\,\text{M}\Omega$, $\alpha = 0{\cdot}975$, $\beta = 39$.

Ans. $$-\frac{110(1 + 5s \times 10^{-6})}{(1 + 29s \times 10^{-6})}$$

1.5 Obtain the transfer function V_o/V_i for the two computing circuits shown in Fig. 1.33. Assume that the operational amplifiers are ideal.

Ans. $$\frac{1}{-s\,2C_1R(1 + sCR/2)}, \quad -\frac{1}{s^2 4(RC)^2}$$

1.6 The effect of eddy currents induced in the iron of the magnetic circuit of a d.c. machine can be approximated by the equivalent circuit shown in Fig. 1.34. Making the assumption that there is

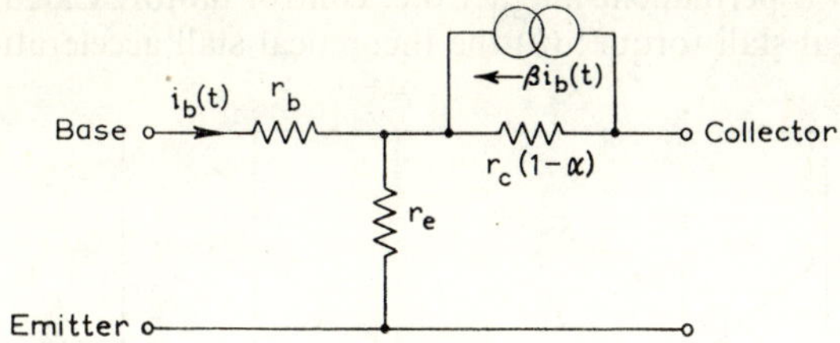

FIG 1.32

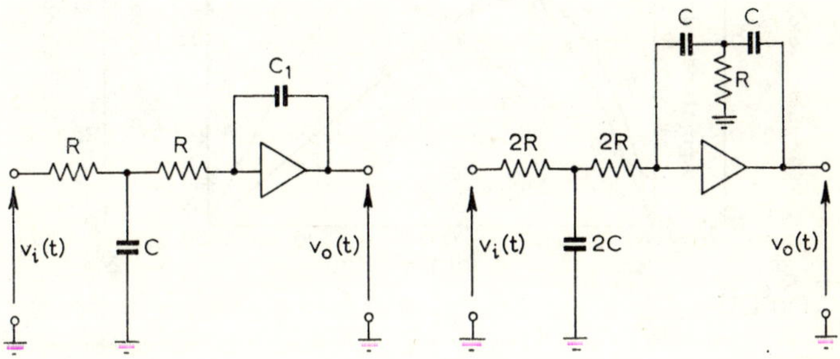

FIG 1.33

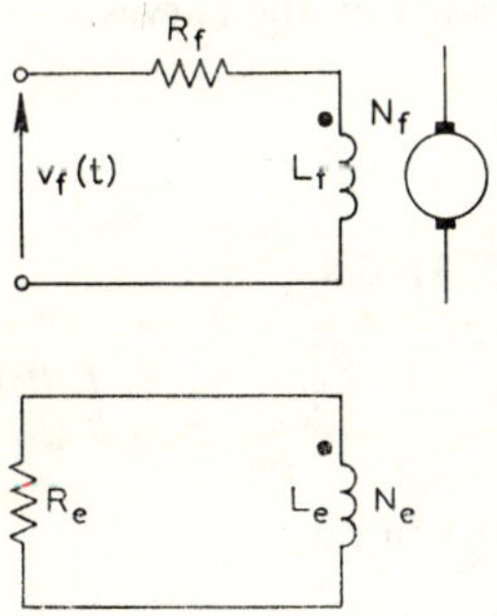

FIG 1.34

perfect coupling between the windings, show that the relation between e.m.f. and input voltage has the form

$$\frac{E}{V_f} = \frac{KN_f}{R_f(1 + s(\tau_e + \tau_f))}$$

where K gives the relation between machine e.m.f. and ampere-turns on the frame.

1.7 The performance curves and data in Fig. 1.35 are supplied in connection with a permanent-magnet d.c. control motor. Calculate (*a*) the theoretical stall torque; (*b*) the theoretical stall acceleration;

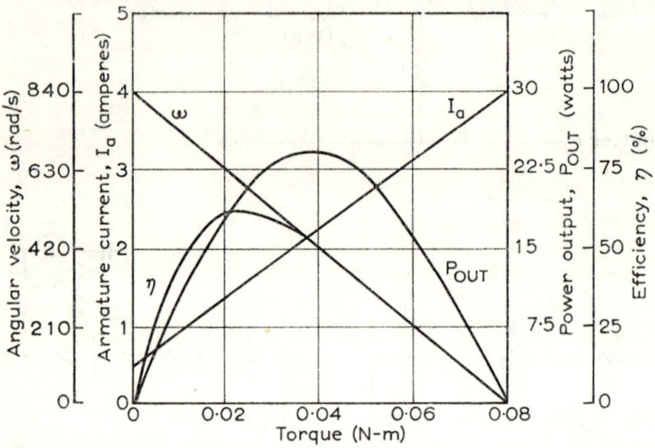

FIG 1.35

(*c*) the damping constant; (*d*) the time-constant. Neglect the effects of armature reaction and friction. Comment on results (*a*) and (*b*) in relation to the actual stall torque indicated by the curves.

Supply voltage, 24 V
Rated output, 23 W
Current at rated output, 1·8 A
Angular velocity at rated output, 523 rad/s
Moment of inertia, $1{\cdot}5 \times 10^{-6}$ kg-m^2

(*I.E.R.E.*)*

Ans. (*a*) 0·1 N-m, (*b*) $6{\cdot}66 \times 10^4$ rad/s^2,
(*c*) $0{\cdot}955 \times 10^{-4}$ N-m/rad-s^{-1}, (*d*) 15·7 ms

1.8 The "diamond" field circuit arrangement is often used to excite large d.c. generators. The advantage of this arrangement is that each separate winding is required to produce 50 % of the ampere-turns on the machine, whereas if reversal were required with each field winding excited separately 100 % ampere-turns would have to be provided.

The circuit shown in Fig. 1.36 represents a push-pull amplifier having zero output impedance feeding the field windings. Obtain the transfer function relating current in a field winding to an amplifier output voltage swing V.

Ans. $\dfrac{1}{R(1 + s(L + M)/R)}$

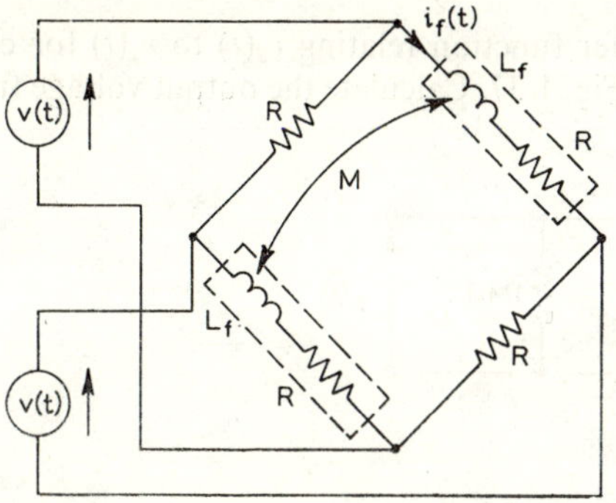

FIG 1.36

1.9 The following data relate to an amplidyne:

Rating, 2 kW
Quadrature circuit resistance, 4 Ω
Quadrature circuit inductance, 1·2 H
Control-field winding resistance, 15 Ω
Control-field winding inductance, 1·8 H
Quadrature e.m.f. for 1 A in control field, 220 V
Output e.m.f. for 1 A in quadrature circuit, 110 V
Output circuit resistance, 3·8 Ω
Output circuit inductance, negligible

If the machine supplies a load of 25 Ω, determine the transfer function relating current in the load to the voltage applied to the control field circuit.

Ans. $\dfrac{14}{(1 + 0{\cdot}12s)(1 + 0{\cdot}3s)}$

1.10 For the d.c. servo motor of Example 1.12, calculate the mechanical time-constant using the following information:

Rotor inertia, $1{\cdot}28 \times 10^{-4}$ kg-m^2
Armature resistance, 0·6 Ω
E.M.F., $5{\cdot}25 \times 10^{-2}$ V/rad-s^{-1}

If the motor is connected to a load having a moment of inertia of 1·85 kg-m^2 by a 100:1 step-down gearbox, find the new time-constant. Determine the gearbox ratio required for maximum acceleration of the motor.

Ans. 28 ms, 69 ms, 120.

1.11 Determine the transfer function relating $v_2(t)$ to $v_1(t)$ for each of the networks shown in Fig. 1.37. Calculate the output voltage from

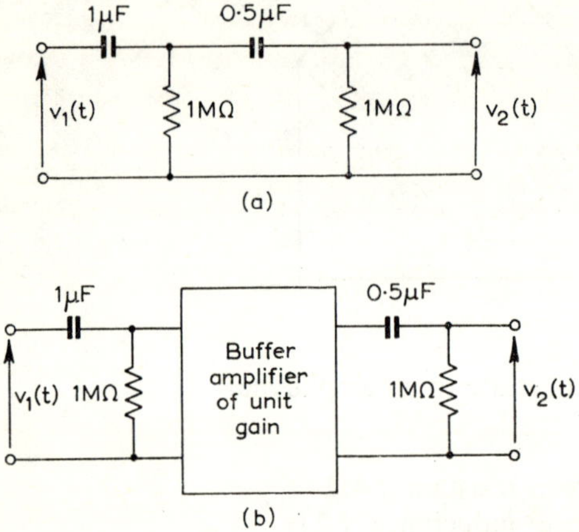

FIG 1.37

each network, t ⩾ 0, for a unit step voltage input at $t = 0$.

(*Southampton Univ.*)

Ans. $\frac{s^2}{s^2 + 4s + 2}$, $\frac{s^2}{2 + 3s + 2}$,

$1{\cdot}2e^{-3{\cdot}4t} - 0{\cdot}21e^{-0{\cdot}59t}$, $2e^{-2t} - e^{-t}$.

1.12 Determine the closed-loop transfer function for the feedback unit shown in Fig. 1.38, assuming that A represents a large negative gain and the amplifier input and output impedances are infinite and negligible respectively. Describe briefly with diagrams the use of such a feedback unit. (*Manchester Univ.*)

Ans. $$\frac{-[s^2C_1C_2R_2R_3 + s(C_1R_2 + C_2R_3 + C_2R_2) + 1]}{sC_2R_1}$$

1.13 What do you understa d by a linear system? Fig. 1.39 shows a linear system in which a 2-phase servomotor drives a load through a gear train. The motor torque/speed characteristics are approximated by parallel straight lines, as shown in the diagram, and the following constants are given:

$f = 49 \times 10^{-3}$ N-m/rad-s^{-1}
Total moment of inertia referred to load shaft, 10^{-3} kg-m^2
Gear ratio, 1:30
Find the transfer function $\theta(s)/V(s)$, where $\theta(t)$ is the angle of the output shaft. If $v(t)$ is a step function, sketch $\theta(t)$ as a function of t. (*Rugby*)*

Ans. $$\frac{0{\cdot}21}{s(1 + 0{\cdot}0046s)}$$

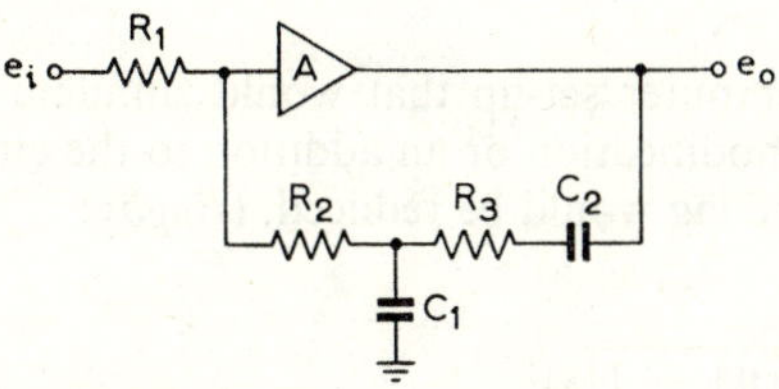

FIG 1.38

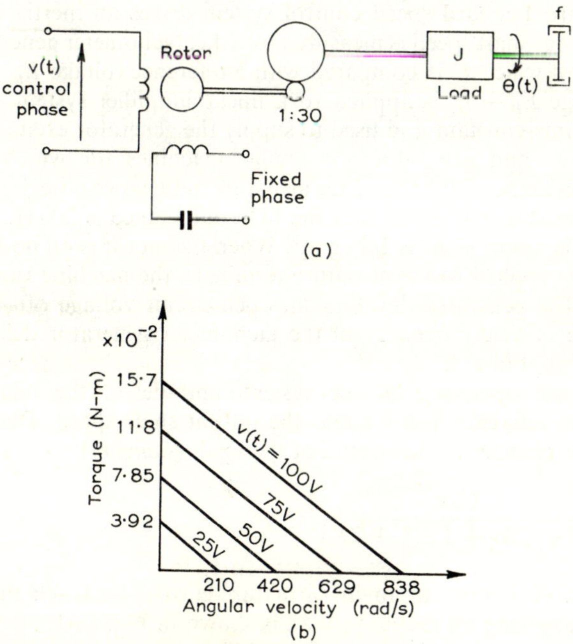

FIG 1.39

1.14 Fig. 1.40 shows an equivalent circuit representation of a two-stage rotary amplifier. The following relations are given: $v_i = 200i_1$, $v_2 = 200i_2$.

Find an expression for the transfer function $v_o(s)/v_i(s)$.

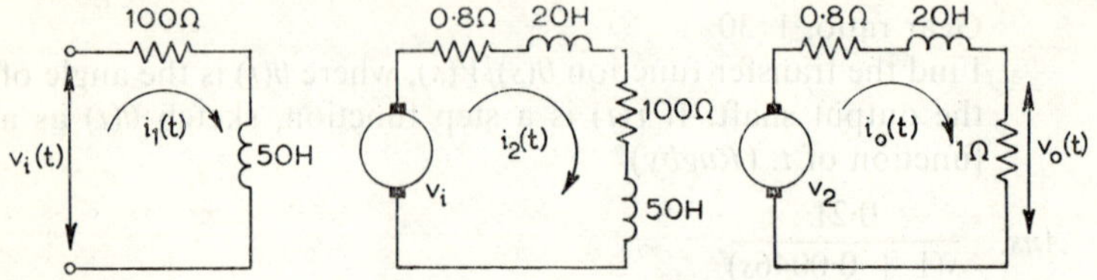

FIG 1.40

Establish an analogue computer set-up that would simulate this rotary amplifier. Suggest a modification or an addition to the circuit of Fig. 1.40 whereby the time lag would be reduced. (*Rugby*)

Ans. $$\frac{2{\cdot}2}{(1 + 0{\cdot}5s)(1 + 0{\cdot}7s)(1 + 11s)}$$

1.15 A Ward Leonard speed control system drives an inertia load and the output shaft speed is measured by a d.c. tachometer generator the output of which E_t is compared with a reference voltage E_R. The error voltage $E_R - E_t$ is applied to a linear amplifier system with negligible time-constant and used to supply the generator excitation.

The motor and generator are similar machines for which the armature resistance is 0·9 Ω and the armature inductance is negligible. The field resistance is 200 Ω and the field inductance is 100 H. The inertia at the motor shaft is 1·2 kg-m². When the motor is on no-load and 240 V is applied to the armature terminals, the machine runs at 210 rad/s. The generator develops an open-circuit voltage of 480 V per ampere of field current, and the tachometer generator delivers $1{\cdot}05 \times 10^{-2}$V/rad-s^{-1}.

Explain the operation of the system, and derive the relation between the reference voltage and the output shaft speed. Discuss the effect of changes in the inertia of the load. (*Bangor*)*

Ans. $$\frac{2{\cdot}5K}{s^2 + 2{\cdot}6s + 1{\cdot}2(1 + 2K)}$$

1.16 A d.c. generator employing armature-current feedback to the field and supplying an inductive load is shown in Fig. 1.41.

Derive the system transfer function $i_g(s)/e_f(s)$.

Determine the effect of the current feedback on the system performance, given the following numerical values:

$$R_f = 1{\cdot}1\,\Omega \qquad R_g = 0{\cdot}2\,\Omega \qquad R = 0{\cdot}4\,\Omega$$
$$L_f = 0{\cdot}25\,\text{H} \qquad L_g = 0{\cdot}01\,\text{H} \qquad L = 1{\cdot}49\,\text{H}$$
$$K_V = 20{\cdot}4\,\text{V/A} \qquad R_o = 0{\cdot}4\,\Omega$$

Sketch the response of the load current to a step change in $e_f(t)$ with and without the feedback.

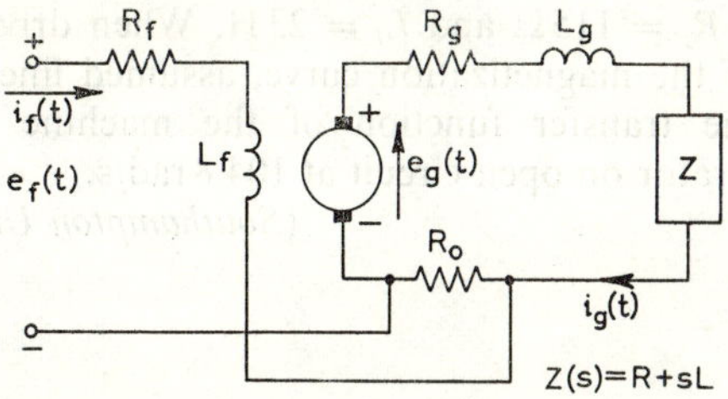

FIG 1.41

Give a practical example of a system in which the technique would be used to advantage. (*Birmingham Univ.*)

Ans. $\dfrac{2{\cdot}2}{1 + 0{\cdot}2s + 0{\cdot}04s^2}$

1.17 A d.c. servomotor has its split field separately excited from a push-pull amplifier. The armature is connected in series with an external resistance to a constant supply voltage. Taking the base or reference value of speed as that corresponding to zero torque, full voltage and rated field m.m.f., show that the speed in per unit values is given by the expression

$$\omega = \frac{k_V - R}{k_V{}^2}$$

for a constant-torque load, and

$$\omega = \frac{k_V(1 - R)}{k_V{}^2(1 - R) + R}$$

for a load having an ideal viscous friction characteristic. In the above expressions, all quantities are per unit; the torque per ampere k_V and the armature circuit resistance R being based respectively on rated m.m.f. and the ratio of the supply voltage to full-load current.

If it is required that the speed/field-current characteristic should always have a positive slope up to full speed, i.e. up to rated field m.m.f., show that the minimum value of R must not be less than 0·5 per unit in each case. (*Manchester Univ.*)

1.18 Derive the transfer function for a constant-speed separately excited d.c. generator operating on open-circuit. Show how it may be represented in a block diagram form suitable for use with an analogue computer. Obtain an expression for the instantaneous armature current supplied to a load, having resistance and inductance, connected to the generator.

A d.c. machine has $R_f = 115\,\Omega$ and $L_f = 23\,\text{H}$. When driven at 52·4 rad/s, the slope of the magnetization curve, assumed linear, is 275 V/A. Calculate the transfer function of the machine as a separately-excited generator on open-circuit at 104·8 rad/s.

(*Southampton Univ.*)*

Ans. $\dfrac{4{\cdot}78}{1 + 0{\cdot}2s}$

1.19 A metadyne generator has the following parameters:

Control field (variator) resistance, 25 Ω
Control field (variator) inductance, 2·5 H
Direct-axis armature resistance, 3 Ω
Direct-axis armature inductance may be neglected.
Quadrature-axis armature resistance, 3 Ω
Quadrature-axis armature inductance, 0·75 H

The machine is run at constant speed and loaded by a 30 Ω resistor; under these conditions its characteristics are as follows:

Load voltage, 300 V
Quadrature-axis steady-state generated voltage, 480 V per field ampere and 120 V per direct-axis armature ampere.
Direct-axis steady-state generated voltage, 120 V per quadrature-axis armature ampere.

Develop the transfer function relating load current to control voltage, and calculate the power gain. Neglect non-linearities and magnetic leakage. (*I.E.R.E.*)

Ans. $\dfrac{25{\cdot}6}{(1 + 0{\cdot}25s)(1 + 0{\cdot}1s) + 160}$ amperes per volt, 19·2

1.20 A d.c. motor, used for position control, has a rotor inertia J and a viscous friction coefficient F. The field winding has resistance R_f and inductance L_f, and the armature resistance is R_a. Determine the transfer function (output angle/input voltage) of the motor for each of the undernoted conditions:

(i) A constant current is supplied to the armature and the motor is controlled by a voltage applied to the field winding—assume that the rotor torque is proportional to the field current.
(ii) The field excitation is held constant and the control voltage is applied to the armature.

Discuss briefly the scope of application of these two forms of control. (*Scottish Assoc.*)

Ans. $$\frac{K}{sR_fF(1 + sL_f/R_f)(1 + sJ/F)}, \frac{K_T}{s(K_TK_V + R_AF)(1 + s\tau)}$$

$$\text{where } \tau = \frac{R_AJ}{K_TK_V + R_AF}$$

1.21 The network shown in Fig. 1.42 has a second-order transfer function when the emitter follower has unit gain, infinite input

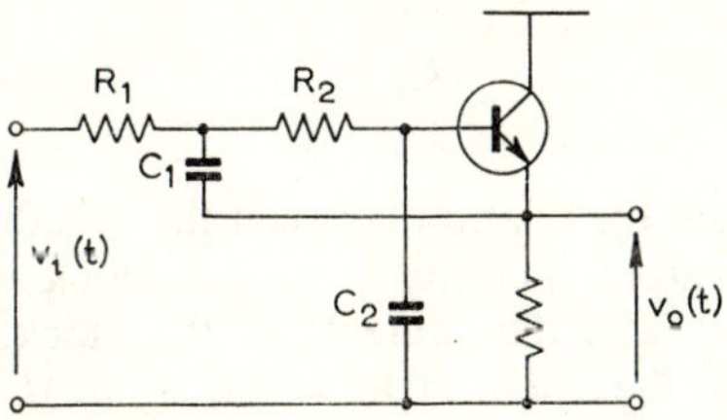

FIG 1.42

impedance and zero output impedance. Derive this transfer function $V_o(s)/V_i(s)$ in terms of R_1, R_2, C_1 and C_2.

The poles of the transfer function are to be on a circle of radius $10\,000\,s^{-1}$ centred on the origin of the *s*-plane, and at an angle of 60° measured from the negative real axis. If $R_1 = R_2 = 1\,k\Omega$, what are the required values of C_1 and C_2? (*Edinburgh Univ.*)

Ans. $$\frac{1}{1 + sC_2(R_1 + R_2) + s^2R_1R_2C_1C_2}$$

$$C_1 = 0{\cdot}2\,\mu F, \quad C_2 = 0{\cdot}05\,\mu F.$$

Chapter 2

Block Diagrams: Signal-flow Graphs: Analogue Computation

It is now obvious that large complicated systems have to be represented by many simultaneous differential equations. In this chapter we introduce two graphical methods of obtaining the required overall transfer functions, namely the block diagram and the signal-flow graph—the latter method becoming more useful as the complexity of the problem increases.

Block Diagrams

In the block diagram approach the relation between the variables is put into a block, and the output quantity from the block is the product of the input quantity and the contents of the block. For example, in Chapter 1 we met the equation $V_f = (R_f + sL_f)I_f$. The block diagram for this is shown in Fig. 2.1 (*a*), and an equally valid alternative in Fig. 2.1 (*b*).

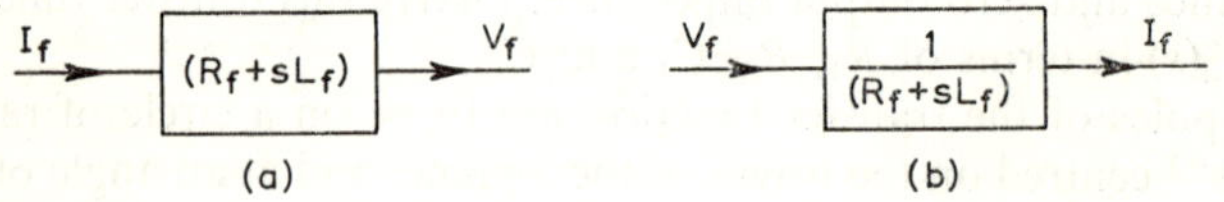

FIG 2.1 The content of the block is the transfer function. Arrows indicate the flow of information.

If the components of a system are cascaded in such a way that interaction does not take place, i.e. so that each can be described independently by a transfer function, the block diagrams too can be cascaded. In Example 1.9 the system was described by the equations

$$I_f = V_f/R_f(1 + s\tau_f) \qquad \text{and} \qquad E = KI_f$$

This is shown in block diagram form in Fig. 2.2 (*a*).

If we have no particular interest in an intermediate variable such as I_f, we can suppress it by combining the contents of the blocks by multiplication as in Fig. 2.2 (*b*). Inserting the values given in Example 1.9 we obtain the transfer function

$$\frac{E}{V_f} = \frac{7{\cdot}7}{1 + 2{\cdot}5s}$$

and the appropriate block diagram is shown in Fig. 2.2 (*c*).

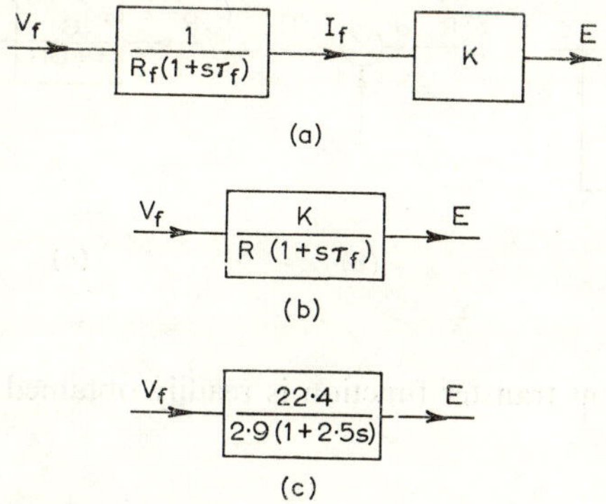

FIG 2.2

Signal-flow Graphs

The signal-flow graph consists of a network in which nodes representing each of the system variables are connected by directed branches. A branch acts as a one-way signal multiplier, the ratio of output to input being defined as the *transmittance*, and an arrowhead is used to indicate the flow of information as in the block diagram.

For the simple examples given above the signal-flow graphs are shown in Fig. 2.3.

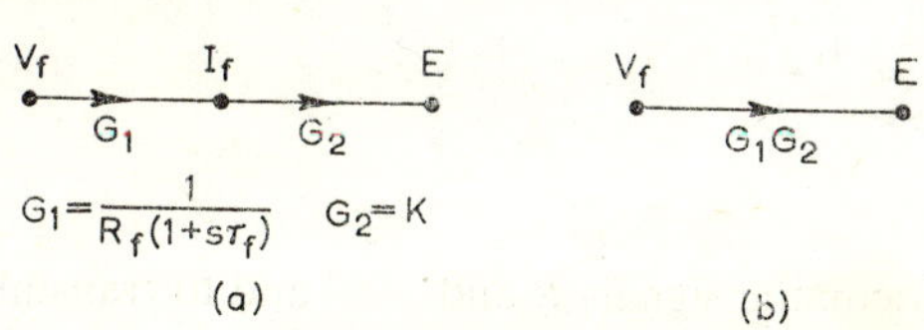

FIG 2.3

The Feedback Connection

The feedback condition is given in block diagram form in Fig. 2.4: (*a*) shows negative feedback and (*b*) shows the amendment for positive feedback.

To obtain the relation between C and R in Fig. 2.4 (*a*) we write down the simultaneous equations involved:

$$C = GE \tag{2.1}$$

$$C' = HC \tag{2.2}$$

$$E = R - C' \tag{2.3}$$

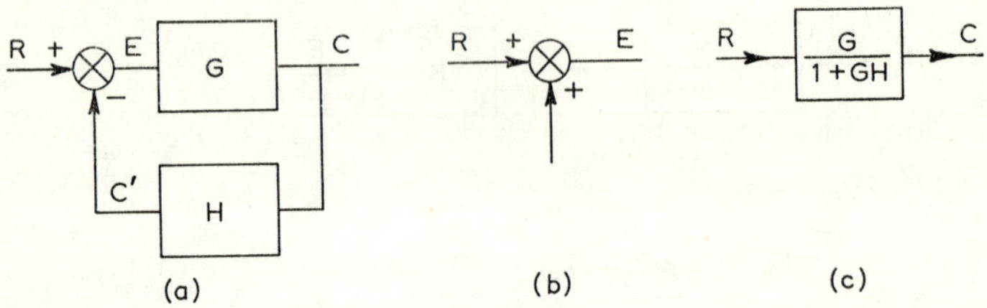

FIG 2.4

The required closed-loop transfer function is readily obtained as

$$C = \frac{G}{1 + GH} R$$

This is shown in block diagram form in Fig. 2.4 (*c*). For positive feedback, eqn. (2.3) becomes $E = R + C'$, giving

$$C = \frac{G}{1 - GH} R$$

We now use the signal-flow diagram to describe the set of equations (2.1), (2.2) and (2.3). This is shown in Fig. 2.5. Here the node labelled

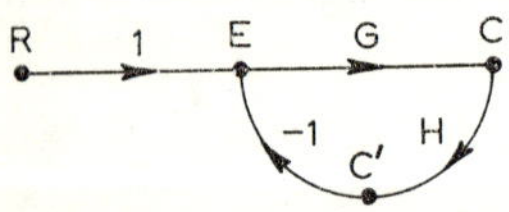

FIG 2.5

E serves to add the incoming signals R and $-C'$ and to transmit the total node signal $(R - C')$ to the outgoing branch.

A node such as R is called a *source node* as it only has outgoing branches. A node such as C (Fig. 2.6 (*c*)) is called a *sink node* as it has only incoming branches. A *common node* has both incoming and outgoing branches. A *path* is any connected sequence of branches whose arrows are in the same direction, e.g. there is a path between R and C involving branches 1 and G.

Fig. 2.5 can be simplified as shown in Fig. 2.6. Fig. 2.6 (*a*) is obtained simply by removing node C', and Fig. 2.6 (*b*), by removing node E. We note that there is still a path between R and C having the transmittance G and a self-loop of $-GH$. This can be represented as in Fig. 2.6 (*c*) as is shown by writing the equations represented by Fig. 2.6 (*b*):

$$C = RG + (-GH)C$$

and

$$\frac{C}{R} = \frac{G}{1 + GH}$$

Mason's Rule

A method which can be used when determining an overall transfer function from the signal-flow diagram is *Mason's rule.* This states that the overall transmittance T is given by

$$T = \frac{\Sigma T_n \Delta_n}{\Delta}$$

T_n is the transmittance of the nth forward path between the source node and the sink node, and Δ is the determinant of the graph and is given by

$$\Delta = 1 - \Sigma T_{n1} + \Sigma T_{n2} - \Sigma T_{n3} + \ldots$$

where T_{n1} = Transmittance of each closed path or loop
T_{n2} = Product of the transmittances of two non-touching loops (i.e. having no nodes in common)

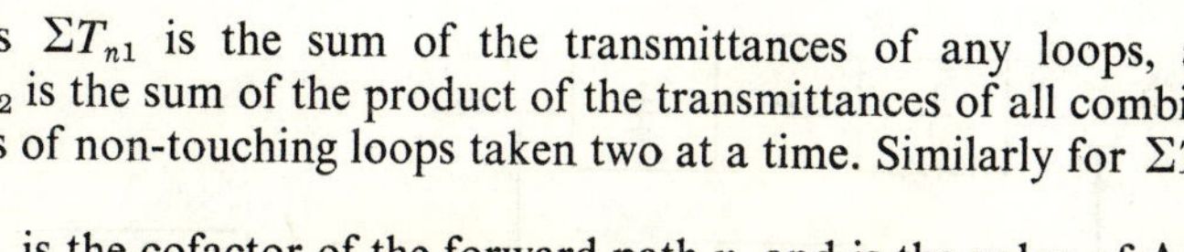

FIG 2.6

Thus ΣT_{n1} is the sum of the transmittances of any loops, and ΣT_{n2} is the sum of the product of the transmittances of all combinations of non-touching loops taken two at a time. Similarly for ΣT_{n3}, etc.

Δ_n is the cofactor of the forward path n, and is the value of Δ for that part of the graph which does not touch the path n.

The use of Mason's rule is illustrated in Examples 2.6 and 2.7, and the rule is used again in Example 3.10.

Analogue Computing

From the preceding discussion it is apparent that different types of system (e.g. electrical, mechanical, thermal) can give rise to the same block diagram or signal-flow graph. If this happens one system can be thought of as being an analogue of the other. Often a complex system can be investigated by means of its analogue, which may be easier to build and to use for measurements than the real system. A very convenient analogue is an electrical one, and this can be obtained with the aid of the operational amplifier.

The operational amplifier has been introduced in Chapter 1; by suitable choice of computing elements it can be used to perform the functions of summation, multiplication by a constant, and integration. There is also the advantage that such amplifiers can be interconnected without undesired interaction taking place.

Operational amplifiers can be interconnected to produce the same block diagrams as the system under investigation. Alternatively, they can be interconnected so that the differential equations governing their output voltages are identical with those describing the variables of the real system. The former method of using an analogue computer is described as *simulation.*

Example 2.1

A rotor having a moment of inertia J (kg-m^2) is coupled to a viscous damper requiring a torque of B (N-m/rad-s^{-1}). Draw a block diagram relating the rotor angular velocity ω (rad/s) to the applied torque T (N-m). Hence obtain the transfer function relating these variables.

Solution

The equation of motion is $T = J\alpha + B\omega$, where α is the angular acceleration (rad/s^2). Hence $\alpha = s\omega$. The block diagram is shown in Fig. 2.7 and in simplified form in Fig. 2.8.

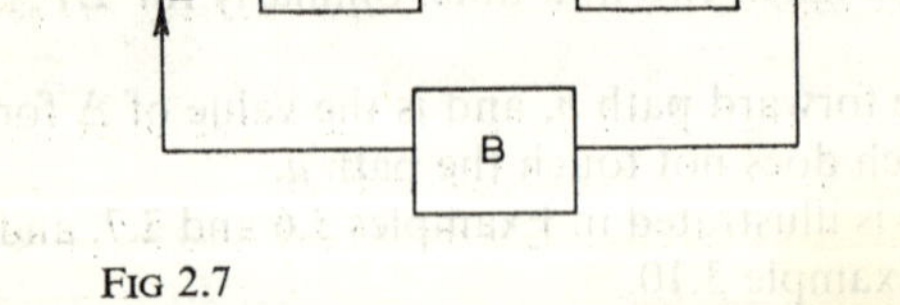

FIG 2.7

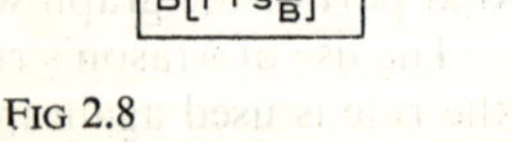

FIG 2.8

Example 2.2

Fig. 2.9 shows a typical arrangement by which the field current of an electric motor may be controlled. Draw a block diagram of the system

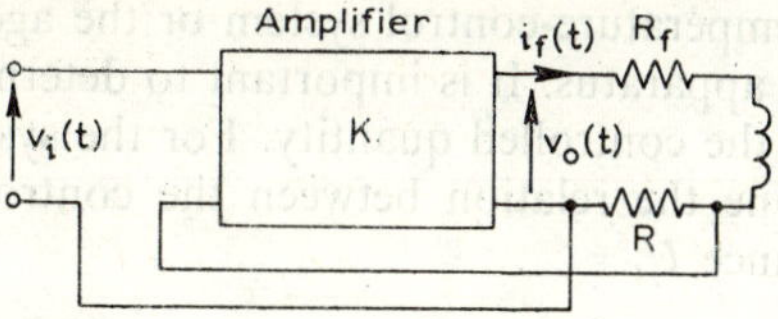

FIG 2.9

and simplify to obtain the relation between V_i and I_f. By comparison with the final block diagram and that for the field circuit alone obtain the factor by which the time-constant has been reduced.

Solution
The block diagram of the system is shown in Fig. 2.10. This can be reduced as shown in Fig. 2.11. Comparison with the original block

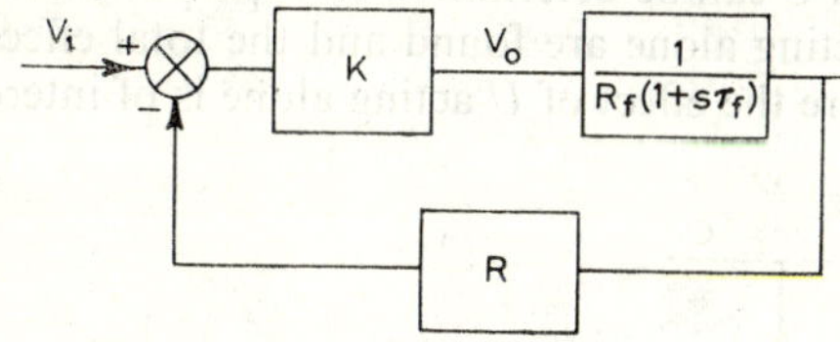

FIG 2.10

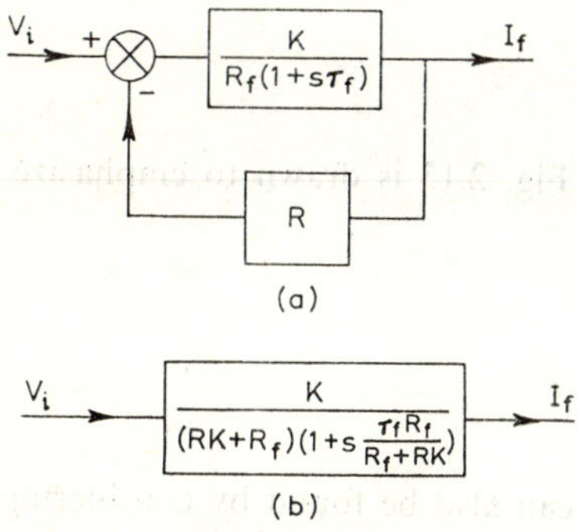

FIG 2.11

diagram shows that the time-constant τ_f has been multiplied by a factor $R_f/(R_f + RK)$.

Example 2.3

A control system is often subjected to unwanted signals or disturbances. These may take the form, for example, of a change in the ambient temperature in a temperature-control system or the ageing of components in electronic apparatus. It is important to determine the effect of disturbance on the controlled quantity. For the system shown in Fig. 2.12 determine the relation between the controlled quantity C and the disturbance U.

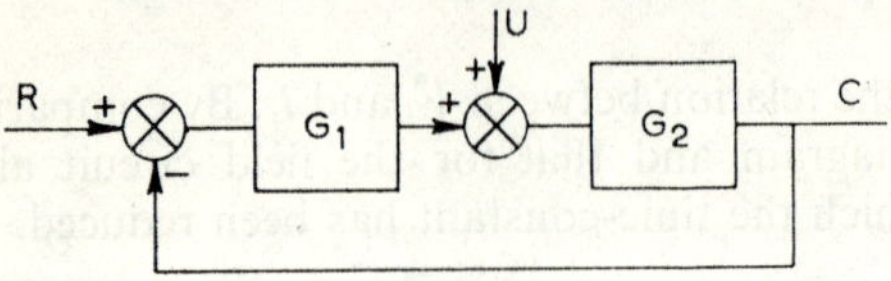

FIG 2.12

Solution

The effects of R and U on C can be determined by superposition, i.e. the effects of R and U acting alone are found and the total effect is obtained by addition. Here the effect of U acting alone is of interest.

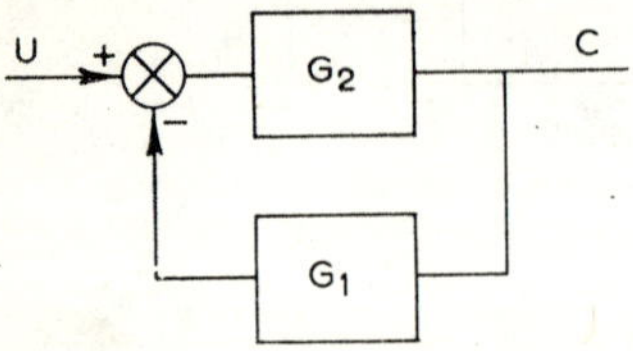

FIG 2.13

The block diagram shown in Fig. 2.13 is drawn to emphasize this effect. Thus

$$\frac{C}{U} = \frac{G_2}{1 + G_1 G_2}$$

Alternative Solution

The effect of the disturbance can also be found by considering the equivalent block diagram, Fig. 2.14. Then the relation

$$\frac{C}{U/G_1} = \frac{G_1 G_2}{1 + G_1 G_2}$$

is easily obtained. Hence

$$\frac{C}{U} = \frac{G_2}{1 + G_1G_2}$$

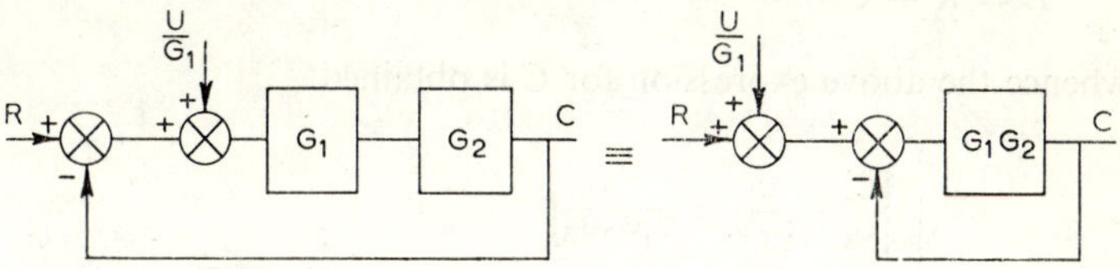

FIG 2.14

Example 2.4

In the block diagram shown in Fig. 2.15, U_1, U_2, U_3 represent disturbances. Determine the relation between the controlled quantity C and the reference quantity R together with the disturbances.

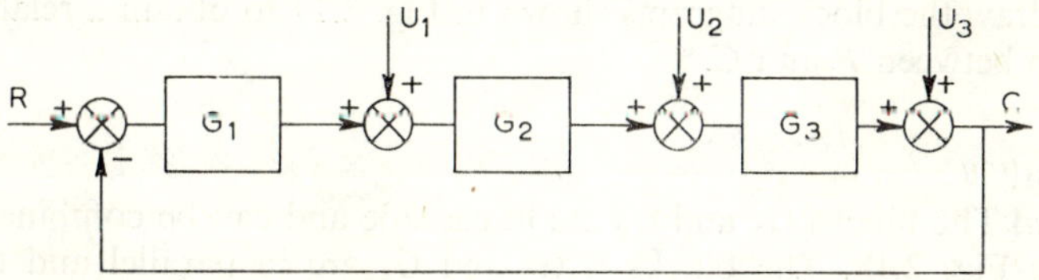

FIG 2.15

Solution

Each of these disturbances can be moved back block by block until they are all present together with R at the input to the system. Transferring a disturbance from the output of a block to its input involves dividing it by the expression in the block. The system can therefore be represented as in Fig. 2.16. Hence

$$C = \frac{G_1G_2G_3}{1 + G_1G_2G_3}\left[R + \frac{U_1}{G_1} + \frac{U_2}{G_1G_2} + \frac{U_3}{G_1G_2G_3}\right]$$

$$= \frac{G_1G_2G_3}{1 + G_1G_2G_3} R + \frac{G_2G_3}{1 + G_1G_2G_3} U_1$$

$$+ \frac{G_3}{1 + G_1G_2G_3} U_2 + \frac{1}{1 + G_1G_2G_3} U_3$$

An alternative method is to write down the simultaneous equations described by the original block diagram:

$$C = U_3 + G_3U_2 + G_2G_3U_1 + G_1G_2G_3E$$
$$E = R - C$$

whence the above expression for C is obtained.

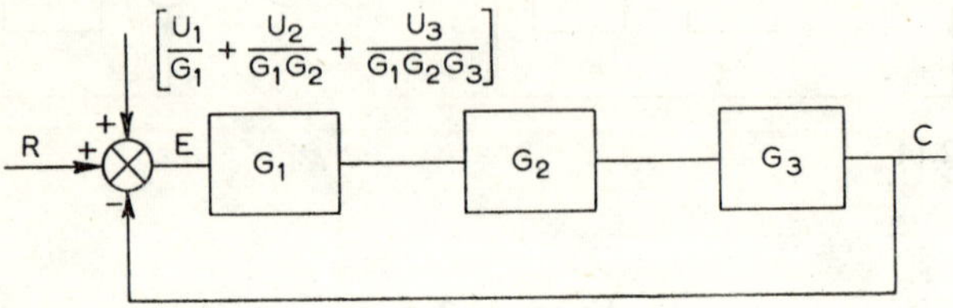

FIG 2.16

Example 2.5

Redraw the block diagrams shown in Fig. 2.17 to obtain a relationship between R and C.

Solution

(*a*) The blocks G_1 and G_2 are in cascade and can be combined to give Fig. 2.18. The blocks G_1G_2 and G_4 are in parallel and their transfer functions can be added. The overall system is now in the form of two blocks in cascade and can be combined to give the relationship between R and C. These steps are shown in Fig. 2.19.

(*b*) The summation point following the G_2 block can be moved forward, which involves rewriting the G_4 block as G_4G_3. This is shown in Fig. 2.20. The arrangement of the three blocks G_2, G_3, G_5 is identical with that of G_1, G_2, G_4 in part (*a*); they can be reduced giving the form shown in Fig. 2.21 (*a*). The diagram then takes the form of two blocks in parallel which can be combined to give Fig. 2.21 (*b*).

(*c*) The summation point for G_4 can be moved back to a point between G_1 and G_2. This entails rewriting the G_4 block as G_4/G_2. The resultant diagram is shown in Fig. 2.22. The diagram can now be simplified as in Fig. 2.23.

(*d*) The diagram is first simplified by reduction of the inner feedback loop; this gives Fig. 2.24. Combination of the cascade blocks gives Fig. 2.25 (*a*), which can then be reduced to Fig. 2.25 (*b*).

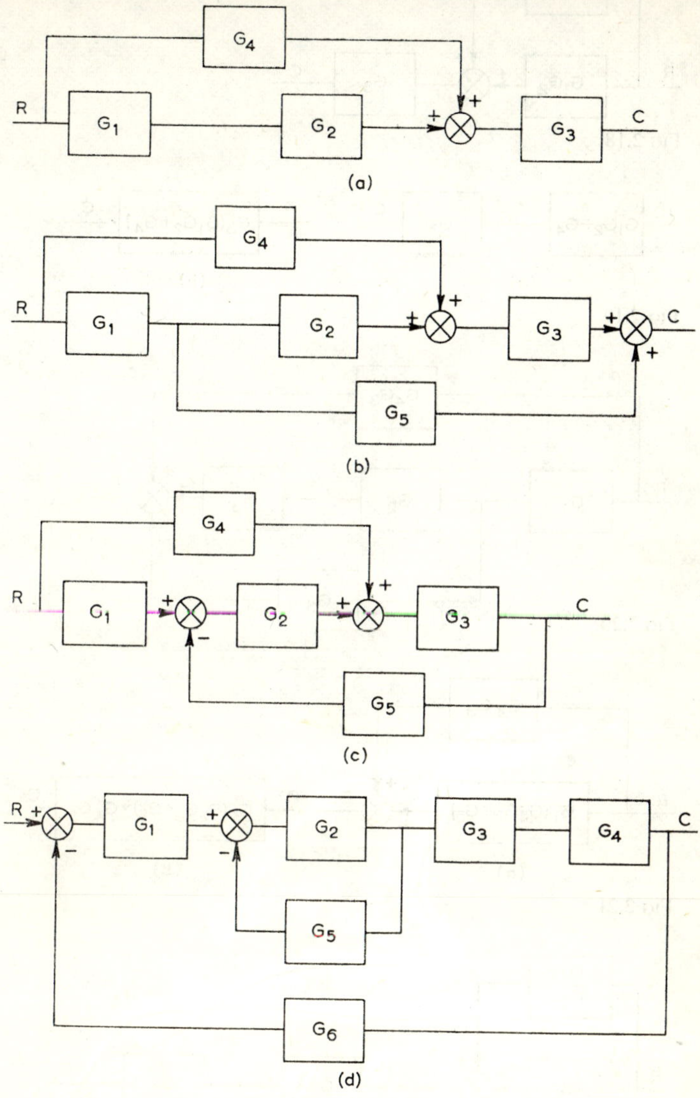

G4
R
G1
G2
G3
C
(a)
G4
R
G1
G2
G3
C
G5
(b)
G4
R
G1
G2
G3
C
G5
(c)
R
G1
G2
G3
G4
C
G5
G6
(d)

FIG 2.17

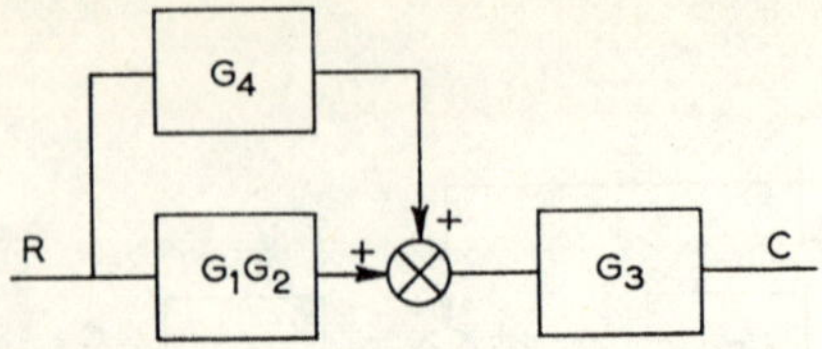

FIG 2.18

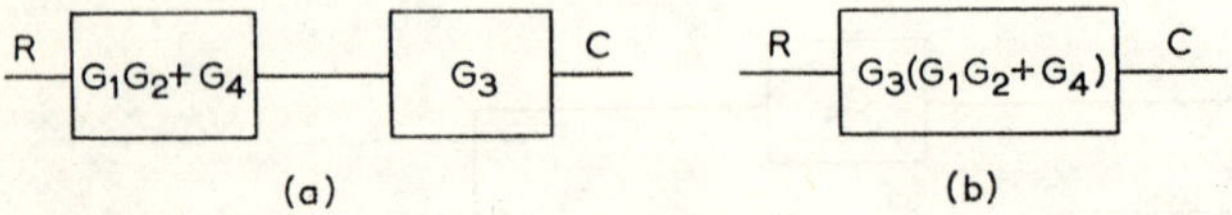

FIG 2.19

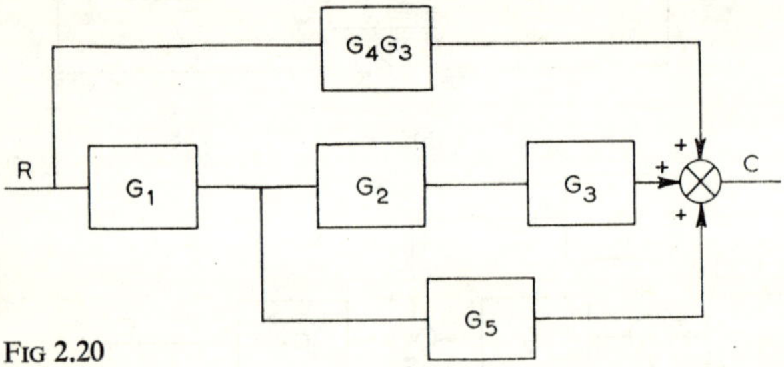

FIG 2.20

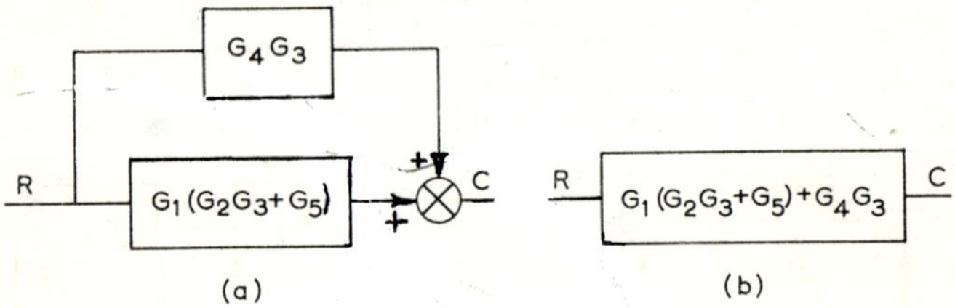

FIG 2.21

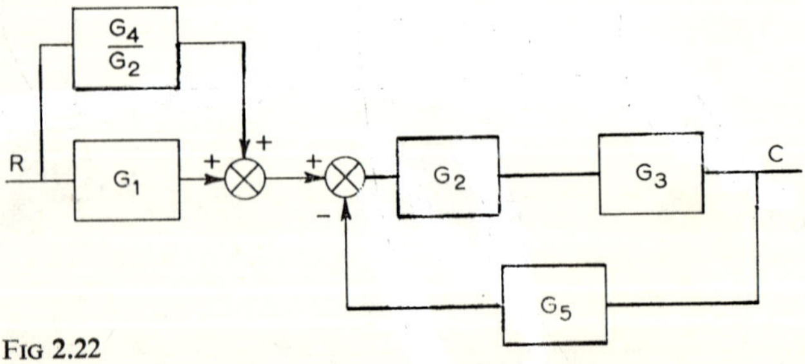

FIG 2.22

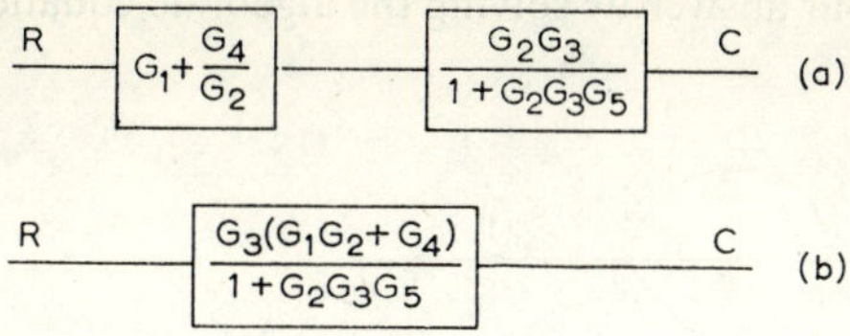

FIG 2.23

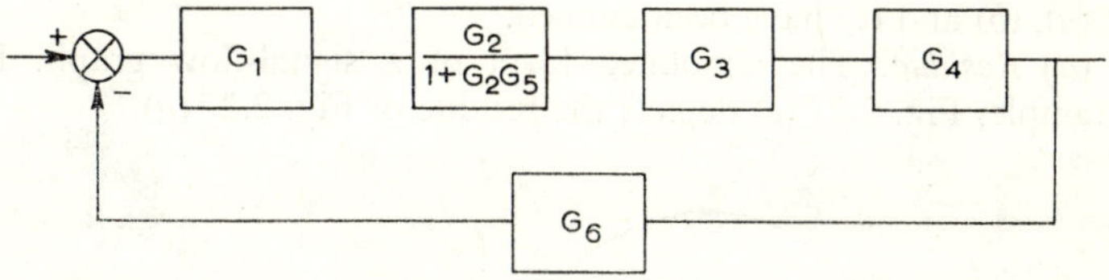

FIG 2.24

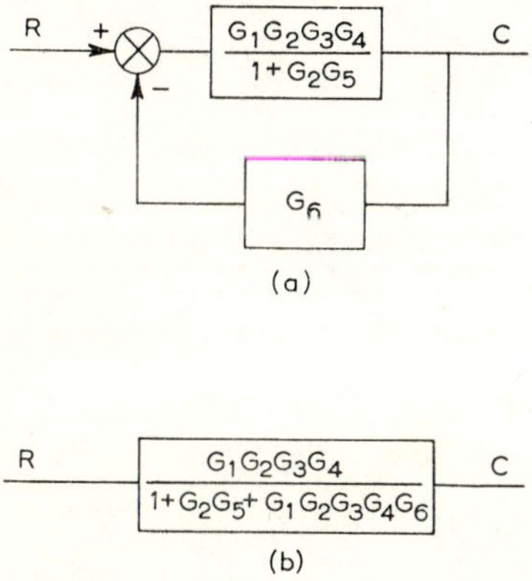

FIG 2.25

Example 2.6

Explain briefly the following terms as used in signal-flow graph analysis: (*a*) common node, (*b*) branch, (*c*) transmission, (*d*) residue, (*e*) index, (*f*) feedback path.

Write down the algebraic equations represented by the flow graph in Fig. 2.26. By simplifying the flow graph, find the transmission from x_1 to x_5. Verify your answer by solving the algebraic equations (*Manchester Univ.*)

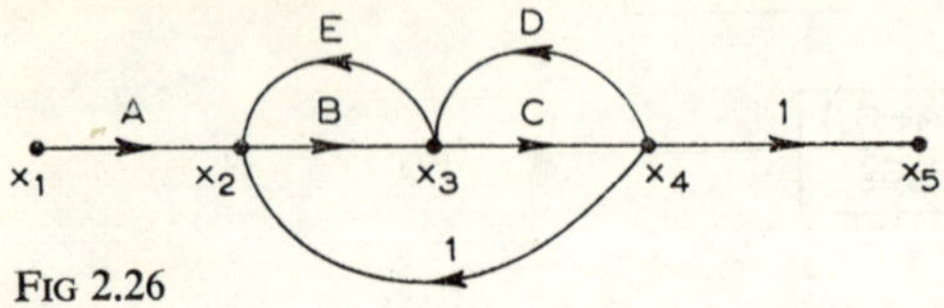

FIG 2.26

Solution

(*a*), (*b*) and (*c*) have been defined.

(*d*) *Residue*. The simplified form of a signal-flow graph. For example, Fig. 2.27 (*e*) depicts the residue of Fig. 2.27 (*a*).

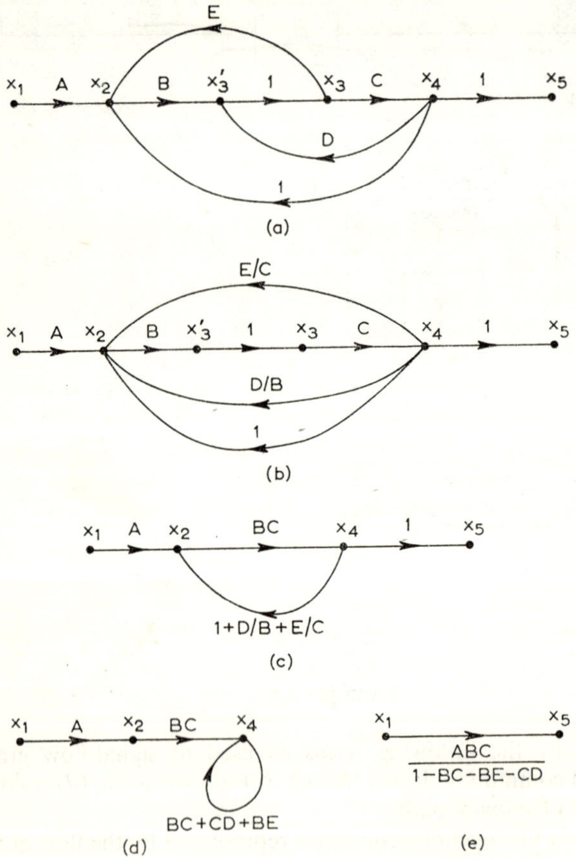

FIG 2.27

(*e*) *Index*. The minimum number of node splittings that are necessary to open all feedback loops.

(*f*) *Feedback path*. Set of branches making up a closed-loop path.

In simplifying the graph of Fig. 2.26 care must be taken at the common node x_3. We proceed by adding a subsidiary node x_3'. The simplification is shown in Fig. 2.27. The algebraic equations are

$$x_2 = Ax_1 + Ex_3 + x_4 \qquad x_3 = Bx_2 + Dx_4$$
$$x_4 = Cx_3 \qquad x_5 = x_4$$

It is left to the reader to verify the relation between x_5 and x_1, indicated by Fig. 2.27 (*e*). This relation can be obtained by use of Mason's rule. Here we have three loops having transmittances *BE*, *CD* and *BC*. There are no non-touching loops. In addition, there is only one forward path between nodes x_1 and x_5 having transmittance *ABC*. Hence the determinant of the graph Δ is given by

$$\Delta = 1 - \Sigma T_{n1} = 1 - (BE + CD + BC)$$
$$\Delta_1 = 1 - 0$$

there being no loops remaining when the forward path is removed from the graph. Therefore

$$\frac{x_5}{x_1} = \frac{ABC}{1 - (BE + BC + CD)}$$

Example 2.7

A linear process with two inputs and two outputs may be represented by the interconnection of transfer functions shown within the dotted lines in Fig. 2.28. This process is to be made non-interacting by the introduction of feedback through the transfer functions C_{12} and C_{21} so that y_1 depends on $f(e_1)$ only and y_2 depends on $f(e_2)$ only. e_1 and e_2 represent the departure of the plant outputs y_1 and y_2 from their respective desired values. Show that

$$C_{12} = \frac{-P_{12}}{P_{11}P_{22} - P_{12}P_{21}} \qquad \text{and} \qquad C_{21} = \frac{-P_{21}}{P_{11}P_{22} - P_{12}P_{21}}.$$

Discuss the difficulties which may arise in connection with the synthesis of non-interacting controllers in general. (*Manchester Univ.*)

Solution

The signal-flow graph for the system of Fig. 2.28 is shown in Fig. 2.29. As we are interested in the interaction between y_1 and $f(e_1)$ and

y_2 and $f(e_2)$, we can find these relations by applying Mason's rule to the manipulated flow diagram of Fig. 2.30. In this diagram we

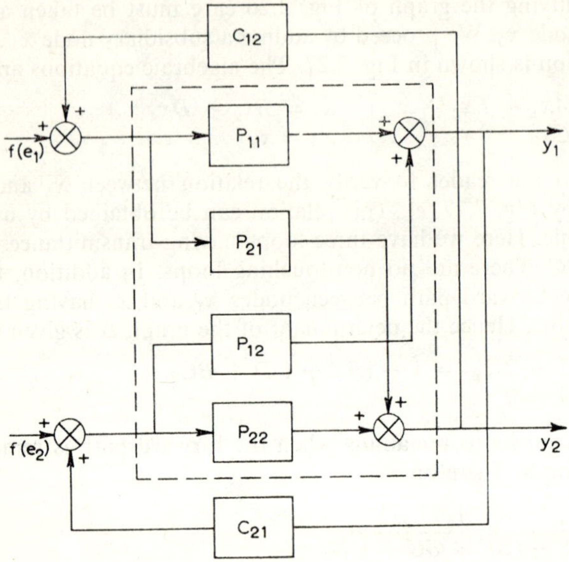

FIG 2.28

emphasize the relation between $f(e_1)$ and y_2 and have opened it out to illustrate the loops involved.

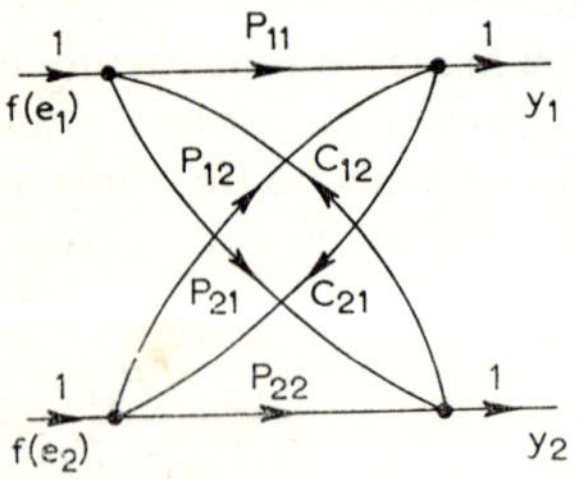

FIG 2.29

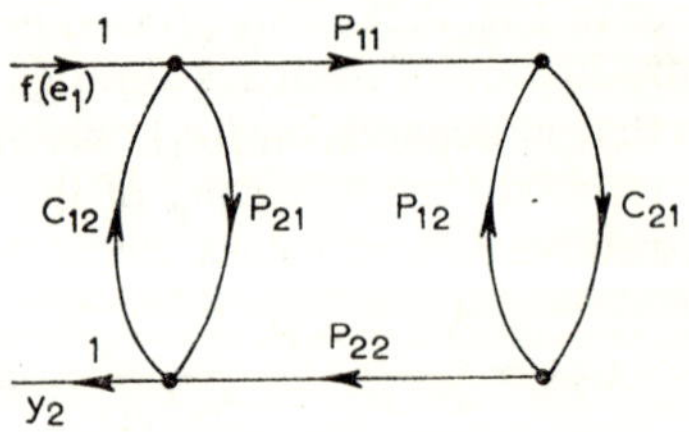

FIG 2.30

The transmittances of the forward paths are P_{21} and $P_{11}C_{21}P_{22}$. The corresponding cofactors are $1 - (P_{12}C_{21})$ and 1. Therefore

$$\frac{y_2}{f(e_1)} = \frac{P_{21}(1 - P_{12}C_{21}) + P_{11}C_{21}P_{22}}{\Delta}$$

For $f(e_1)$ to have no effect on y_2 this expression must be zero. Hence

$$C_{21} = \frac{-P_{21}}{P_{11}P_{22} - P_{12}P_{21}}$$

By symmetry,

$$C_{12} = \frac{-P_{12}}{P_{11}P_{22} - P_{12}P_{21}}$$

This problem could be solved directly from the block diagram, the equations which must be simplified being

$$y_1 = P_{11}s + P_{12}t \qquad \text{and} \qquad y_2 = P_{22}t + P_{21}s$$

where new variables s and t have been defined:

$$s = f(e_1) + C_{12}y_2 \qquad t = f(e_2) + C_{21}y_2$$

Example 2.8

Show from first principles that an operational amplifier may be used to perform the operations of summing, integration and multiplication by a constant coefficient.

Use a block diagram to show how the above operations may be used to solve a differential equation of the form

$$a\frac{d^2y}{dt^2} + b\frac{dy}{dt} + cy + k = 0$$

Assume that at $t = 0$, $dy/dt = 0$ and $y = 0$. (*I.E.R.E.*)

Solution

The general circuit of an operational amplifier has already been considered in Example 1.8, where it was shown that

$$\frac{V_o}{V_i} = -\frac{Z_2(s)}{Z_1(s)}$$

In this expression $Z_2(s)$ is the operational impedance of the feedback element, and $Z_1(s)$ the operational impedance of the input element.

If Z_1 and Z_2 are both pure resistances, R_1 and R_2, then

$$v_o(t) = -\frac{R_2}{R_1}v_i(t)$$

and the amplifier has performed the function of multiplying by a constant coefficient R_2/R_1.

For $Z_1 = R_1$ and $Z_2 = 1/sC$ (i.e. the feedback element is a pure capacitor C), then

$$V_o = -\frac{V_i}{sCR}$$

Converting to the time domain,

$$v_o(t) = -\frac{1}{CR}\int_0^t v_i(t)\,dt$$

This gives the function of integration with initial condition $v_o = 0$ at $t = 0$.

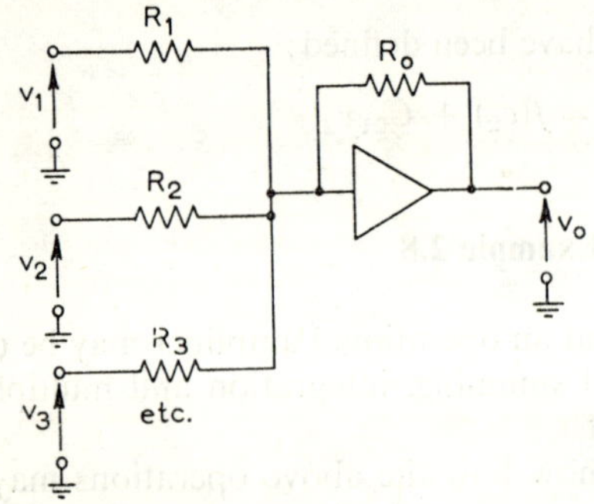

FIG 2.31

To obtain summation the circuit shown in Fig. 2.31 is used. Summation of currents at the virtual earth point gives

$$\frac{V_1}{R_1} + \frac{V_2}{R_2} + \frac{V_3}{R_3} + \ldots + \frac{V_o}{R_o} = 0$$

$$V_o = -R_o\left(\frac{V_1}{R_1} + \frac{V_2}{R_2} + \frac{V_3}{R_3} + \ldots\right)$$

If $R_o = R_1 = R_2 = R_3 = \ldots$, this gives summation; otherwise it gives the combined function of scaling (multiplication by a constant) and summation.

The differential equation to be solved can be written in the form

$$\frac{d^2y}{dt^2} = -\left(\frac{b}{a}\frac{dy}{dt} + \frac{c}{a}y + \frac{k}{a}\right)$$

If it is assumed that a voltage is available equal to d^2y/dt^2, two integrations will give voltages proportional to dy/dt and y. These voltages can then be summed in the correct proportions to give the right-hand side of the above equation. Making the sum of these voltages equal to $-d^2y/dt^2$ gives the required configuration, which is shown in Fig. 2.32.

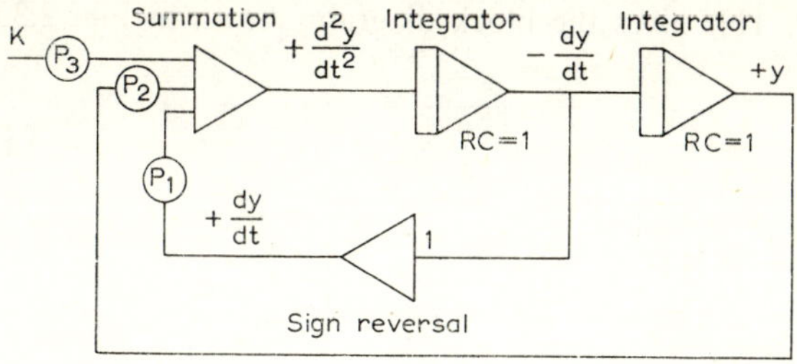

FIG 2.32

P_1, P_2 and P_3 represent voltage dividers. These dividers and the summing resistors are set so that dy/dt is added with a coefficient b/a; y is added with a coefficient c/a; and the constant voltage k is added with a coefficient $1/a$.

Example 2.9

Draw a block diagram of an analogue computer to solve the equations

$$\frac{d^2x_1}{dt^2} + \frac{dx_1}{dt} + (x_1 - x_2) = f(t)$$

$$\frac{d^2x_2}{dt^2} + \frac{dx_2}{dt} + (x_2 - x_1) = 0$$

Explain how initial conditions may be taken into account.

What is the minimum number of integrators that may be used in a system to represent the equations? (*I.E.R.E.*)

Solution

The method of obtaining the block diagram is similar to that used in the previous example. The equations are rewritten

$$\frac{d^2x_1}{dt^2} = -\left[\frac{dx_1}{dt} + (x_1 - x_2) - f(t)\right]$$

$$\frac{d^2x_2}{dt^2} = -\left[\frac{dx_2}{dt} + (x_2 - x_1)\right]$$

Voltages are assumed to exist that are proportional to d^2x_1/dt^2 and d^2x_2/dt^2. Two integrations of each of these voltages will give voltages proportional to dx_1/dt, x_1, dx_2/dt and x_2. Summation of the appropriate variables will give voltages that can be made equal to d^2x_1/dt^2 and d^2x_2/dt^2. This gives the block diagram shown in Fig. 2.33. In

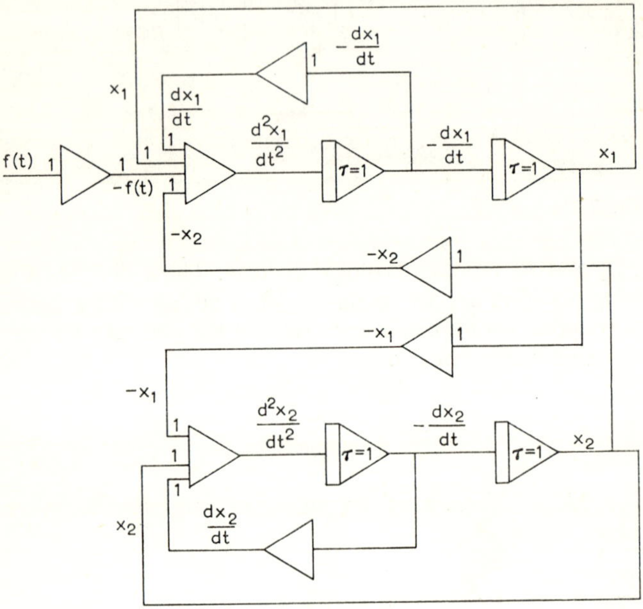

FIG 2.33

this diagram τ represents the time-constant (RC) of the integrators. Initial conditions can be taken into account at the integrators. If at $t = 0$ the capacitors in the integrators are charged, the integrators will give an output

$$v_o = \frac{1}{RC}\int_0^t v_i\,dt + v_{\text{initial}}$$

where v_{initial} represents initial conditions. Suitable relay switching has to be provided to ensure that the integrators commence their normal integrating action at $t = 0$.

The number of integrations necessary to solve these equations must always be four, but the number of pure integrators required can be reduced by using a summing integrator as shown in Fig. 2.34.

Using the virtual earth principle and summing currents, we obtain

$$\frac{v_1}{R_1} + \frac{v_2}{R_2} + \frac{v_3}{R_3} + \frac{v_o}{R} + C\frac{dv_o}{dt} = 0$$

If $R_1 = R_2 = R_3 = R$, then

$$v_o + CR\frac{dv_o}{dt} = -(v_1 + v_2 + v_3)$$

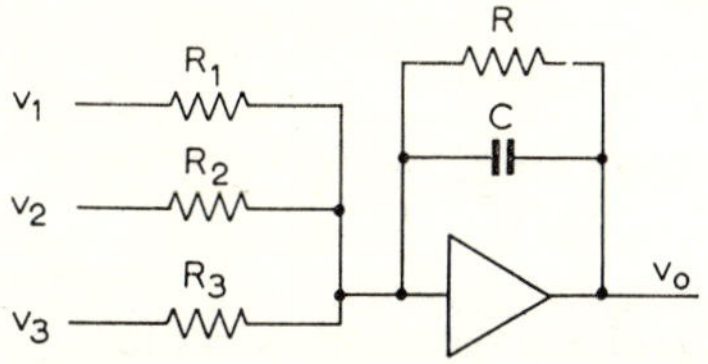

FIG 2.34

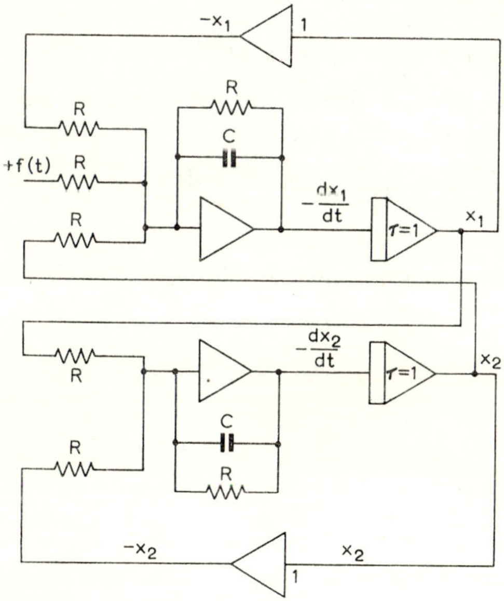

FIG 2.35

Putting $CR = 1$, the equations can be represented by the system shown in Fig. 2.35.

The disadvantage of this system is that voltages representing d^2x_1/dt^2 and d^2x_2/dt^2 are not available.

Example 2.10

An analogue computer is to be used to analyse the transient behaviour of a simple series circuit consisting of a 22 pF capacitor initially charged to 250 V, a 56 μH inductor, and a variable resistor that can take values in the range 1–5 kΩ.

It is required to display the variation with time of the current in the circuit, i, and its derivatives, di/dt and d^2i/dt^2, using a multi-channel pen recorder.

The computer components available are

Operational amplifiers: working range ± 100 V
Computing resistors: 1 MΩ, 100 kΩ
Computing capacitors: 1 μF
Coefficient potentiometers [voltage dividers] as needed.

The specifications of the pen recorder are

Preferred frequency of operation, about 1 Hz;
Paper speed, 20 cm/s;
Deflection sensitivity (all channels), 40 V/cm.

Draw a complete patching diagram for the computer, including numerical values for all coefficients, and give the scale factors applicable to horizontal and vertical deflections on each channel of the pen recorder. (*Southampton Univ.*)

Solution

The circuit to be analysed is shown in Fig. 2.36. The differential equation governing this circuit is

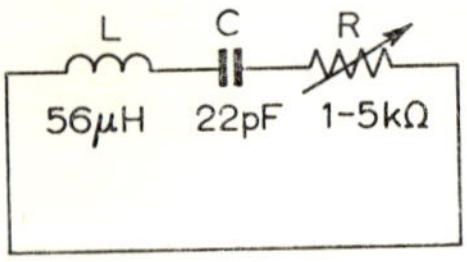

FIG 2.36

$$L\frac{d^2i}{dt^2} + R\frac{di}{dt} + \frac{1}{C}i = 0 \qquad (1)$$

with the appropriate initial conditions for i and di/dt. These initial conditions can be obtained from the equation for the Laplace transform of current, which is

$$I\left(R + sL + \frac{1}{sC}\right) = \frac{V_0}{s}$$

where the right-hand term is due to the initial voltage V_o on the capacitor. Hence

$$I = \frac{V_0}{s\left(R + sL + \dfrac{1}{sC}\right)}$$

The initial value of current can be obtained from the initial-value theorem:

$$\lim_{t\to 0} f(t) = \lim_{s\to\infty} sf(s) = \lim_{s\to\infty} \frac{sV_0}{s\left(R + sL + \dfrac{1}{sC}\right)} = 0$$

Similarly the initial value of di/dt can be found:

$$\lim_{t\to 0} \frac{di}{dt} = \lim_{s\to\infty} s[sI] = \lim_{s\to\infty} \frac{s^2V_0}{s\left(R + sL + \dfrac{1}{sC}\right)} = \frac{V_0}{L}$$

Eqn. (1) can be rewritten

$$\frac{d^2i}{dt^2} + \frac{R}{L}\frac{di}{dt} + \frac{1}{LC}i = 0 \qquad (2)$$

The values of the coefficients are

$$\frac{R}{L} = 17{\cdot}8 \times 10^6 \to 89{\cdot}4 \times 10^6 \qquad \text{and} \qquad \frac{1}{LC} = 813 \times 10^{12}$$

and the initial conditions are $i = 0$ and $di/dt = 4{\cdot}46 \times 10^6$.

In order to obtain the scaling an estimate of the maximum values must be made. This can be done by considering the solution of the equation with the damping term zero, i.e. the worst condition:

$$\frac{d^2i}{dt^2} + \omega_n^2 i = 0$$

where $\omega_n^2 = 1/LC$. This has a solution

$$i = I_0 \sin \omega_n t$$

and the initial condition for di/dt gives $I_0 = 4{\cdot}46 \times 10^6/\omega_n$. Hence

$$i_{max} = \frac{4{\cdot}46 \times 10^6}{\omega_n}$$

$$\left(\frac{di}{dt}\right)_{max} = 4{\cdot}46 \times 10^6$$

$$\left(\frac{d^2i}{dt^2}\right)_{max} = \omega_n 4{\cdot}46 \times 10^6$$

where $\omega_n = 1/\sqrt{(LC)} = 28{\cdot}7 \times 10^6$ rad/s. Substitution yields

$$i_{max} = 0{\cdot}155\text{ A}$$

$$\left(\frac{di}{dt}\right)_{max} = 4{\cdot}46 \times 10^6\text{ A/s}$$

$$\left(\frac{d^2i}{dt^2}\right)_{max} = 128 \times 10^{12}\text{ A/s}^2$$

In order to obtain convenient scaling factors these values are taken as 0·15 A, 5×10^6 A/s, 150×10^{12} A/s^2. The maximum voltage swing available at the output of the computing amplifiers is ± 100 V.

For most efficient use of this voltage swing the scaling must be such that the maximum value of each variable is represented by 100 V on the computer. This gives scaling factors as follows:

$$i \equiv 1{\cdot}5 \times 10^{-3}\text{ A/V}$$

$$\frac{di}{dt} \equiv 5 \times 10^4\text{ As}^{-1}\text{/V}$$

$$\frac{d^2i}{dt^2} \equiv 150 \times 10^{10}\text{ As}^{-2}\text{/V}$$

Eqn. (2) can be rewritten

$$150 \times 10^{10}\left(\frac{d^2i}{dt^2}\bigg/150 \times 10^{10}\right) + 5 \times 10^4\,\frac{R}{L}\left(\frac{di}{dt}\bigg/5 \times 10^4\right) + \frac{1{\cdot}5 \times 10^{-3}}{LC}\left(\frac{i}{1{\cdot}5 \times 10^{-3}}\right) = 0 \qquad (3)$$

However, $[(d^2i/dt^2)/(150 \times 10^{10})]$ is the voltage of the output of an amplifier representing d^2i/dt^2, and this voltage becomes a variable in the computer. It could be denoted by a new symbol, but for a large problem this would mean considerable cross-checking to find which variable on the computer denoted which physical variable. To overcome this the new variable is written $[d^2i/dt^2]$ and this is interpreted as "the voltage in the computer representing the physical variable d^2i/dt^2." Using the same notation for di/dt and i, eqn. (3) becomes

$$\left[\frac{d^2i}{dt^2}\right] + \frac{5 \times 10^4}{150 \times 10^{10}}\,\frac{R}{L}\left[\frac{di}{dt}\right] + \frac{1{\cdot}5 \times 10^{-3}}{150 \times 10^{10}}\,\frac{1}{LC}\,[i] = 0 \qquad (4)$$

and substitution of the values of R, L and C gives

$$\left[\frac{d^2i}{dt^2}\right] + K\left[\frac{di}{dt}\right] + 0{\cdot}813[i] = 0$$

UWE Library Services
Frenchay Campus
Tel: 0117 32 82277

Borrowed Items 08/05/2015 12:59
XXXX5698

tem Title	Due Date
Solutions of roblems in control ngineering	05/06/2015 23:59
Essentials of control ohn Schwarzenbach	05/06/2015 23:59

Indicates items borrowed today
www.uwe.ac.uk/library

where K can be varied from 0·59 to 2·98 as R varies from 1 to 5 kΩ. Once the scaling factors for the variables have been decided, the time-constants of the integrators can be calculated as follows.

The integrator performs the function

$$v_o = \frac{1}{RC}\int_0^t v_i \, dt$$

Taking

$$v_0 = \frac{i}{1{\cdot}5 \times 10^{-3}} \qquad \text{and} \qquad v_i = \frac{di}{dt}\Big/ 5 \times 10^4$$

gives

$$\frac{i}{1{\cdot}5 \times 10^{-3}} = \frac{1}{RC}\int_0^t \left(\frac{di}{dt}\Big/ 5 \times 10^4\right) dt$$

For this to represent

$$i = \int_0^t \frac{di}{dt}\, dt$$

we must make

$$RC - \frac{1{\cdot}5 \times 10^{-3}}{5 \times 10^4} = 3 \times 10^{-8}\,\text{s}$$

Similarly the time-constant for the second integrator becomes

$$RC = \frac{5 \times 10^4}{150 \times 10^{10}} = \frac{1}{30} \times 10^{-6}\,\text{s}$$

However, the natural frequency of the circuit under investigation is $28{\cdot}7 \times 10^6/2\pi = 4{\cdot}56$ MHz, and the preferred frequency of operation for the pen recorder is about 1 Hz. Any transient from such a circuit would be at far too high a frequency to record, as well as being above the operating frequency of the computing amplifiers. A need for time scaling is indicated: the speed of computation should be reduced by a factor of say 5×10^6. The only time-dependent elements in the computer are the integrators, and their operation can be time scaled by increasing the time-constants by a factor 5×10^6. This gives integrator time-constants of 0·15 s and 0·16 s. The complete patching diagram is shown in Fig. 2.37.

The initial condition is

$$\left.\frac{di}{dt}\right|_{t=0} = 4{\cdot}46 \times 10^6\,\text{A/s}$$

which at a scale of $1\,V \equiv 5 \times 10^4\,A/s$ is represented by 89·2 V. (This in practice would probably be obtained from a 100 V supply and a voltage divider set at 89·2% (0·892 p.u.)). The polarity of this supply would be negative as the output of the integrator concerned represents $-di/dt$.

Using the amplitude and time scaling factors and the paper speed and deflection sensitivity of the pen recorder, the scale factors

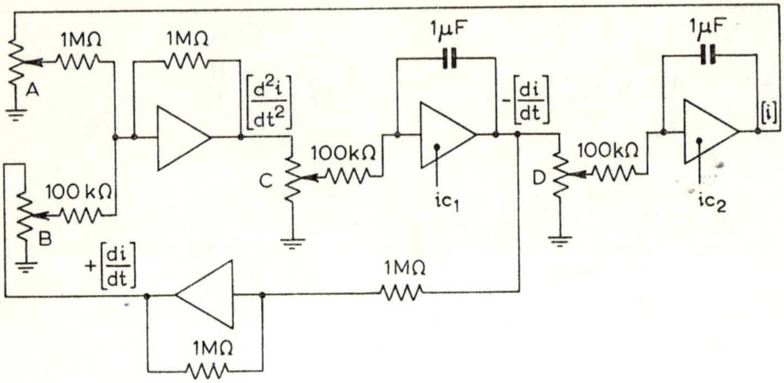

FIG 2.37

Voltage-divider settings in per-unit values:

$A = 0{\cdot}813$
$B = 0{\cdot}059 \rightarrow 0{\cdot}298$ as R varies 1–5 kΩ
$C = 0{\cdot}625$
$D = 0{\cdot}667$

ic_1 = initial condition = −89·2 V
ic_2 = initial condition = 0

applicable to horizontal and vertical deflections can be calculated. These are:

Horizontal: t	$1\,cm \equiv 0{\cdot}25\,ms$
Vertical: i	$1\,cm \equiv 60\,mA$
$\dfrac{di}{dt}$	$1\,cm \equiv 2 \times 10^6\,A/s$
$\dfrac{d^2i}{dt^2}$	$1\,cm \equiv 6 \times 10^{13}\,A/s^2$

PROBLEMS

2.1 A separately-excited 460 V 52·3 rad/s 74·6 kW d.c. motor is directly coupled to a load having an inertia of 106 kg-m². The motor details are: effective inductance of armature circuit, 0·015 H; mean value of armature resistance, 0·1 Ω; inertia of armature, 9·7 kg-m². The resistance and inductance of the supply may be neglected.

Draw a labelled block diagram of the system and hence or otherwise obtain the transfer function relating the speed of the motor to the applied voltage.

Ans. $\dfrac{5}{s^2 + 6{\cdot}6s + 44}$

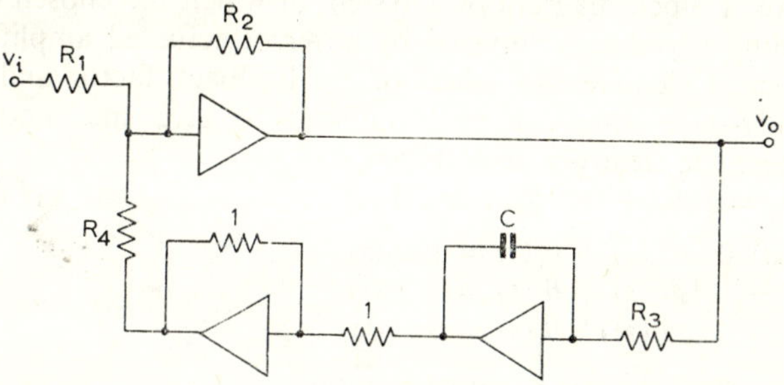

FIG 2.38

2.2 Determine the relation v_o/v_1 for the arrangements of computing amplifiers shown in the diagram of Fig. 2.38. What would be the effect of making R_2 large compared to R_3 and R_4? (*Lanchester*)

$$Ans. \quad \frac{v_o}{v_i}(s) = -\frac{R_4}{R_1}\left[\frac{s\tau}{1 + s\beta\tau}\right]$$

where $\tau = CR_3$, $\beta = R_4/R_2$.

$$\text{As } R_2 \to \infty \quad \frac{v_o}{v_1}(s) \to -s\frac{R_4}{R_1}\tau$$

2.3 The following equations express the output variable θ_o of a plant in terms of three independent variables θ_1, θ_2 and θ_3:

$$\theta_o = \theta_a + \theta_b + \theta_c$$

where

$$0{\cdot}5\frac{d\theta_a}{dt} + \theta_a = 10\theta_1$$

$$4\frac{d\theta_b}{dt} + \theta_b = \theta_2$$

$$\frac{d\theta_c}{dt} + \theta_c = 0{\cdot}1\theta_3$$

The values of θ_1, θ_2, θ_3 lie in ranges of the same order.

(*a*) Assuming all initial values to be zero, give the Laplace transforms of these equations. Using the transfer functions obtained, draw a block diagram of the plant. Which of the three independent variables might be chosen as a controlling variable for the plant, and why?

(*b*) Draw a block diagram of a system in which the chosen controlling variable is adjusted by an error-actuated amplifying device. Calculate the values of the feedback factor and the steady-state amplifier gain required to give the modified system the steady-state equation

$$\theta_o = 10\theta_1 + 0{\cdot}01\theta_2 + 0{\cdot}001\theta_3$$ (*I.E.E.*)

Ans. (*a*)
$$0{\cdot}5s\theta_a(s) + \theta_a(s) = 10\theta_1(s)$$
$$4s\theta_b(s) + \theta_b(s) = \theta_2(s)$$
$$s\theta_c(s) + \theta_c(s) = 0{\cdot}1\theta_3(s)$$

$\theta_o(s)$ is most sensitive to $\theta_1(s)$.

(*b*) Feedback factor, 0·99; amplifier gain, 100.

2.4 Fig. 2.39 illustrates the operation of a vibration table. The

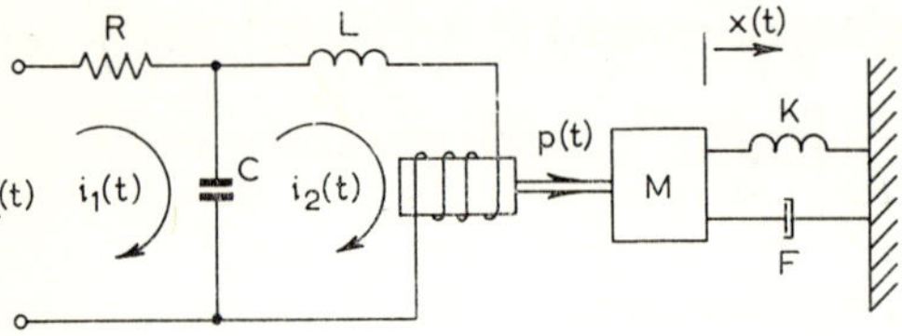

FIG 2.39

following relationships apply: the force $p(t)$ acting on mass $M = k_2 i_2(t)$; the back-e.m.f. of the vibrator coil

$$e_b(t) = k_1 \frac{dx(t)}{dt}$$

k_1 and k_2 are constants; K = spring stiffness and F = viscous friction coefficient. Determine the transfer function $X(s)/E_i(s)$ relating displacement of the mass to the applied voltage. (*Hatfield*)

Ans.
$$\frac{X}{E_i}(s) = \frac{K_2/RK}{a_4s^4 + a_3s^3 + a_2s^2 + a_1s + 1}$$

where
$$a_1 = F/K + L/R + K_1K_2/RK$$
$$a_2 = LC + M/K + FL/KR + K_1K_2C/K$$
$$a_3 = ML/KR + FLC/K$$
$$a_4 = MLC/K$$

2.5 Fig. 2.40 illustrates schematically an electro-mechanical system for controlling the position of a heavy sliding door in a furnace. It comprises two amplifiers of voltage gains A_1 and A_2 and a solenoid actuator of resistance R_a and inductance L_a which positions the door. The door may be represented by a mass M moving in a horizontal direction and restrained by a spring of coefficient k and a damping force of coefficient λ. The position, x, of the door is monitored by means of a rack and pinion, the pinion being coupled via a gearbox to (*a*) a helical potentiometer [voltage divider] for position feedback, and (*b*) a tachometer for feedback compensation, the constants for

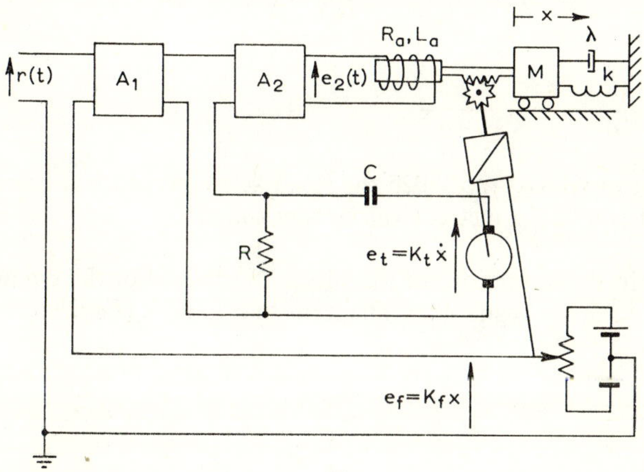

FIG 2.40

the potentiometer and tachometer being K_f and K_t. The reference input is a voltage $r(t)$ with respect to earth, calibrated such that 1 V is equivalent to 0·305 m of door movement.

In the steady state the force p newtons developed by the actuator is numerically equal to 4·48 times the input voltage e_2 volts to the solenoid.

Values for other system constants are:

$M = 36{\cdot}3\,\text{kg}$	$K_f = 3{\cdot}28\,\text{V/m}$
$\lambda = 14{\cdot}7\,\text{N/m-s}^{-1}$	$R = 10\,\Omega$
$k = 14{\cdot}7\,\text{N/m}$	$L = 50\,\text{H}$

A_1, A_2, K_t, R and C are unknowns.

Assuming linear operation throughout, draw the block diagram for the complete system showing clearly the transfer function for each unit.

Show that the transfer function for the closed-loop system is given by

$$\frac{X(s)}{R(s)} = \frac{F_1(s)}{1 + F_2(s)}$$

where

$$F_1(s) = \frac{A}{(s + 0{\cdot}2)(s^2 + 0{\cdot}4s + 0{\cdot}4)}$$

$$F_2(s) = \frac{3{\cdot}28A}{b} \frac{(s^2 + bs + ab)}{(s + a)(s + 0{\cdot}2)(s^2 + 0{\cdot}4s + 0{\cdot}4)}$$

and

$$a = \frac{1}{CR} \qquad b = \frac{3{\cdot}28A_1}{K_t} \qquad A = 0{\cdot}025A_1A_2$$

Show that provided $A \gg 0{\cdot}025$ the steady-state error for a 1 m unit step input is $0{\cdot}025/A$ metre. (*Nottingham Univ.*)*

2.6 Explain the limitations of the Block Diagram. For the circuit in Fig. 2.41 show by Signal-flow Diagram that $i_2 = 0$. (*Lanchester*)

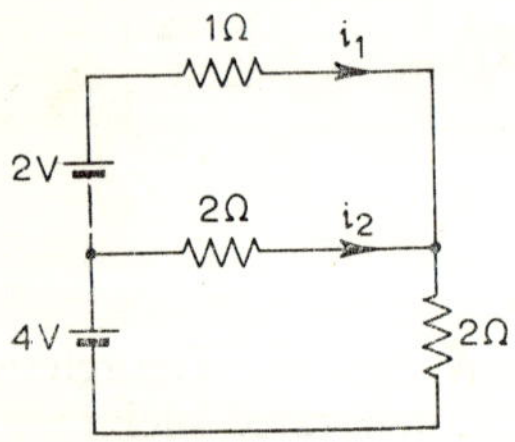

FIG 2.41

2.7 Draw the signal-flow graphs for the system depicted in Fig. 2.17. Obtain the system determinant Δ for each graph and use the results to verify the transfer functions obtained in the solution to Example 2.5.

2.8 Determine the overall closed-loop transfer function of the multi-loop system shown in Fig. 2.42. Also find the characteristic equation for the system and determine the range of values of K for which it will be stable, assuming the transfer functions below:

$H_1 = 5$; $H_2 = K/(1 + s)$; $H_3 = 1/\{s(1 + 0{\cdot}1s)\}$;
$H_4 = 2s$; $H_5 = 3s^2$ (*I.E.R.E.*)

Ans. $$\frac{\theta_o}{\theta_i} = \frac{1}{1/H_1H_2H_3 + H_4/H_1 + H_5 + 1}$$

$$0{\cdot}1s^3 + (15K + 1{\cdot}1)s^2 + (1 + 2K)s + 5K = 0$$

System stable for $K > 0$.

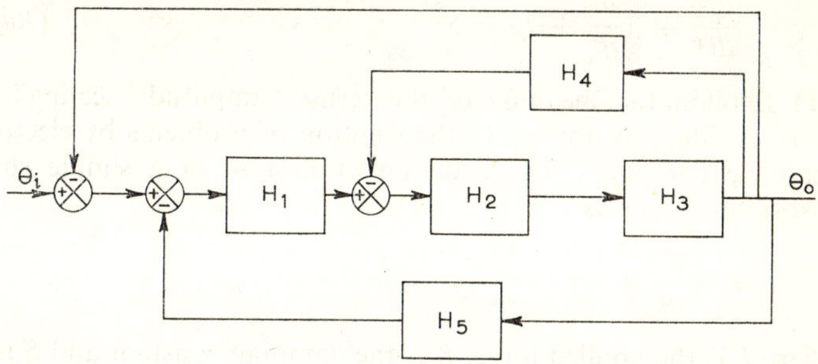

FIG 2.42

2.9 Establish an analogue computer set-up to solve the two simultaneous equations

$$y + 3 = \frac{dx}{dt} \qquad x + 8 = \frac{dy}{dt}$$

In the analogue computer set-up shown in Fig. 2.43, derive an equation relating x to y. If now x is given by the equation $x = 5t$ (t =

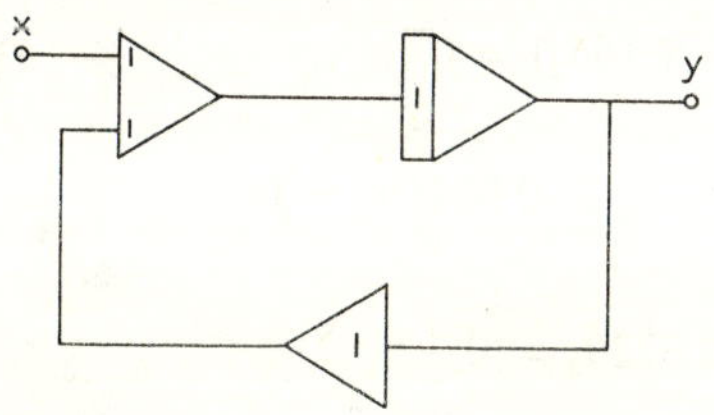

FIG 2.43

time), determine the value of y as a function of time. Sketch this function to a base of time. (*Rugby*)

Ans. $x = y + \dfrac{dy}{dt}$, $\quad y(t) = 5(t - 1 + e^{-t})$

2.10 (*a*) State four properties of an amplifier in order that it may be employed for analogue computation.

(*b*) A sinusoidally varying voltage $V_1 \sin \omega t$ is applied to an analogue integrator. If the output voltage is $V_2 \sin(\omega t + \theta)$, how do V_2 and θ vary with ω? Sketch their variations.

(*c*) Synthesize an analogue set-up to simulate the simultaneous equations

$$u + 4V = \frac{d^2V}{dt^2}$$

$$\frac{d^2u}{dt^2} + 3\frac{du}{dt} + 2u = 5$$

(*Rugby*)

2.11 Explain the meaning of the terms "amplitude scaling" and "time scaling" as applied to the solution of problems by electronic analogue computer. The equation of motion of a simple shock-absorbing system is

$$F = K\frac{dx}{dt} + Sx$$

where F is the applied force, K is the damping constant and S is the spring constant.

If the spring constant can have a maximum value of 750 whilst the applied force and damping constant have values of 12·5 and 250 respectively, determine a suitable computer flow diagram such that the displacement x may be observed in real time. Suggest suitable values for components associated with any amplifiers used. (*Scottish Assoc.*)

2.12 The three simultaneous equations

$$\theta_1 + 0{\cdot}135\frac{d\theta_2}{dt} - 0{\cdot}1(\theta_i - 1{\cdot}15\theta_2) = 0$$

$$0{\cdot}1\frac{d^2\theta_2}{dt^2} + 0{\cdot}109\frac{d\theta_2}{dt} + 1{\cdot}67\theta_1 - 0{\cdot}0011\theta_3 = 0$$

$$0{\cdot}01\frac{d\theta_3}{dt} + 0{\cdot}032\frac{d^2\theta_2}{dt^2} + 0{\cdot}322\theta_2 = 0$$

are to be solved using an analogue computer. Construct the signal-flow graphs for these equations and hence determine a suitable computer programme such that the quantities θ_1, θ_2, θ_3, θ_i can be recorded.

The operational amplifiers can be arranged to act as summers with gains of unity or ten, or integrators having 0·1 or 1 s time-constants. Give some indication of the order of accuracy to be expected in this solution. (*Leeds Univ.*)

2.13 A position control system is shown in block diagram form in Fig. 2.44. Construct a signal-flow graph for the system, and draw up a suitable analogue computer programme for its simulation such that the response to step and ramp function inputs may be investigated as the constants K_1, K_2, K_3 vary between 2–5, 5–20, and 0·5–2,

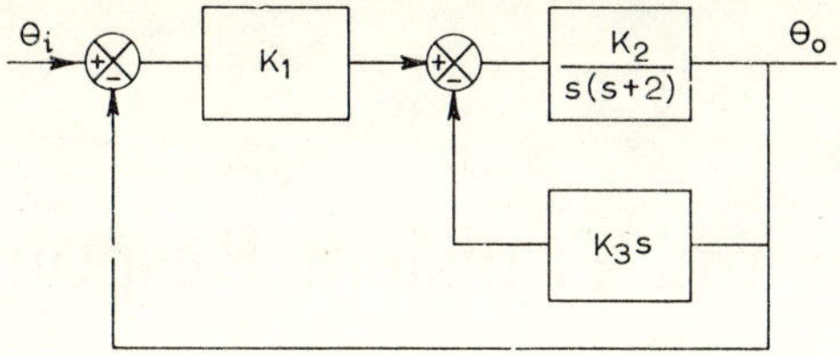

FIG 2.44

respectively. Assume that the operational amplifiers are available as summers with gains of 1 and 10, or as integrators with 0·1 and 1 s time-constants. (*Leeds Univ.*)

2.15 Construct an analogue-computer flow diagram to simulate the multiple-input multiple-output system shown in Fig. 2.45. Assuming

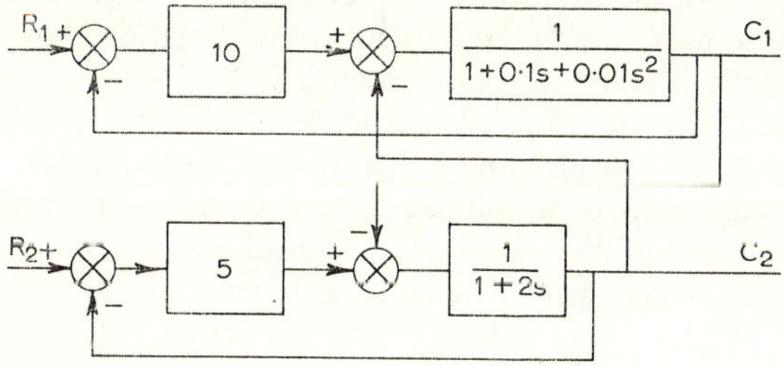

FIG 2.45

that the operational amplifiers have a gain of either 1 or 10, the input resistors are 1 MΩ, and the available capacitors are 0·1 μF and 1 μF. All potentiometer [voltage divider] settings must be shown. (*Lanchester*)

Chapter 3

Steady-state and Transient Response

In the previous chapters we have seen how the transfer functions of the component parts of a system can be obtained, enabling the system to be represented by a block diagram or a signal-flow graph. The overall relation between the various quantities can be obtained by manipulation of such a diagram. In essence the relationship between the quantities is that of an *n*th order differential equation.

Here we deal with the solution to this equation. It is of considerable advantage to make use of the Laplace transformation; however, in simple examples it is by no means essential.

As an introduction to transient response, and for the sake of continuity, we shall consider the generator of Example 1.9. The generator is to be used in a voltage control system. An amplifier having a voltage gain of 50 is available. Determine the proportion K of the generator voltage which must be fed back to the amplifier for the system to have the following open loop gains: 10, 12 and 50. If the amplifier has a time-constant of 50 ms determine the response of the generator voltage for a unit-step-function reference signal for each of these loop gains.

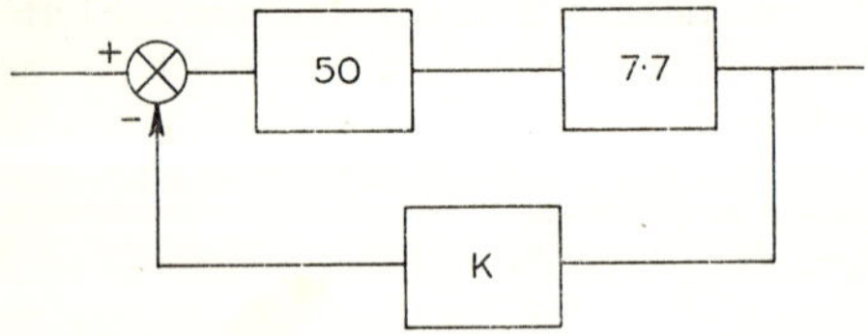

FIG 3.1

A simplified diagram is shown in Fig. 3.1. The open-loop gain is given by $50 \times 7{\cdot}7 \times K$. For a gain of 10,

$$K = \frac{10}{50 \times 7{\cdot}7} = 0{\cdot}026$$

Similarly for gains of 12 and 50, $K = 0{\cdot}031$ and $0{\cdot}13$ respectively.

(*a*) The block diagram is shown in Fig. 3.2 for an open-loop gain of 10. Letting C represent the controlled quantity, here the generator voltage, the closed-loop transfer function is

$$\frac{C}{R}(s) = \frac{50 \times 7{\cdot}7}{(1 + 0{\cdot}05s)(1 + 2{\cdot}5s)} \Big/ \left[1 + \frac{10}{(1 + 0{\cdot}05s)(1 + 2{\cdot}5s)}\right]$$

$$= \frac{3\,080}{(s + 6{\cdot}2)(s + 14{\cdot}2)}$$

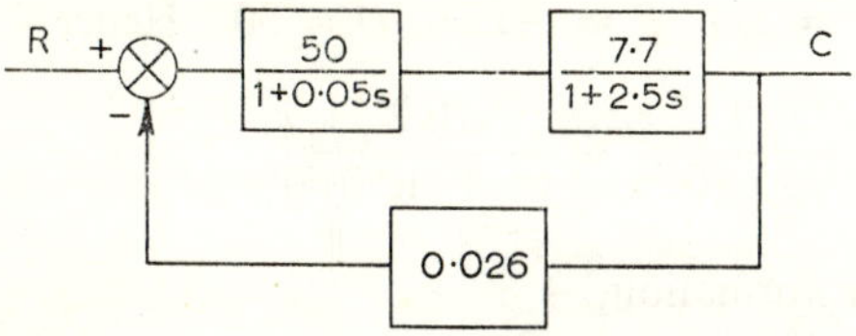

FIG 3.2

For a unity-step-function input, $R(s) = 1/s$. Hence

$$C(s) = \frac{3\,080}{s(s + 6{\cdot}2)(s + 14{\cdot}2)}$$

Splitting this into partial fractions,

$$C(s) = \frac{A}{s} + \frac{B}{(s + 6{\cdot}2)} + \frac{C}{(s + 14{\cdot}2)}$$

where

$$A = \left.\frac{3\,080}{(s + 6{\cdot}2)(s + 14{\cdot}2)}\right|_{s=0} = 35$$

$$B = \left.\frac{3\,080}{s(s + 14{\cdot}2)}\right|_{s=-6{\cdot}2} = -62$$

Similarly $D = 27$. Thus

$$C(s) = \frac{35}{s} - \frac{62}{(s + 6{\cdot}2)} + \frac{27}{(s + 14{\cdot}2)}$$

Converting to the time domain,

$$c(t) = 35[1 - 1{\cdot}78\,\mathrm{e}^{-6{\cdot}2t} + 0{\cdot}78\,\mathrm{e}^{-14{\cdot}2t}] \tag{3.1}$$

(*b*) For an open-loop gain of 12 the closed-loop response is

$$\frac{C}{R}(s) = \frac{3\,080}{s^2 + 20{\cdot}4s + 104} = \frac{3\,080}{(s + 10{\cdot}2)^2}$$

Thus

$$C(s) = \frac{3080}{s(s + 10{\cdot}2)^2}$$

which in partial fraction form becomes

$$C(s) = \frac{A}{s} + \frac{B}{(s + 10{\cdot}2)} + \frac{D}{(s + 10{\cdot}2)^2}$$

where $A = 3080/10{\cdot}2^2 = 29{\cdot}6$, $B = -29{\cdot}6$, $D = 302$. Hence

$$C(s) = 29{\cdot}6\left[\frac{1}{s} - \frac{1}{(s + 10{\cdot}2)} - \frac{10{\cdot}2}{(s + 10{\cdot}2)^2}\right]$$

Taking the inverse transformation,

$$c(t) = 29{\cdot}6[1 - \mathrm{e}^{-10{\cdot}2t} - 10{\cdot}2t\,\mathrm{e}^{-10{\cdot}2t}] \qquad (3.2)$$

(*c*) For an open-loop gain of 50,

$$\frac{C}{R}(s) = \frac{3080}{s^2 + 20{\cdot}4s + 408}$$

Substituting for $R(s)$ and putting into partial-fraction form,

$$C(s) = \frac{A}{s} + \frac{B}{s + 10{\cdot}2 + j17{\cdot}5} + \frac{D}{s + 10{\cdot}2 - j17{\cdot}5}$$

where $A = 3080/408 = 7{\cdot}54$

$$B = 3080/(-10{\cdot}2 - j17{\cdot}5)(-j35) = 4{\cdot}34\underline{/-150^\circ}$$

$$D = 3080/(-10{\cdot}2 + j17{\cdot}5)(j35) = 4{\cdot}34\underline{/150^\circ}$$

Taking the inverse transformation,

$$\begin{aligned} c(t) &= 7{\cdot}54 + 4{\cdot}34\,\mathrm{e}^{-10{\cdot}2t}[\mathrm{e}^{j(17{\cdot}5t+150^\circ)} + \mathrm{e}^{-j(17{\cdot}5t-150^\circ)}] \\ &= 7{\cdot}54 + 8{\cdot}68\,\mathrm{e}^{-10{\cdot}2t}\cos(17{\cdot}5t + 150^\circ) \\ &= 7{\cdot}54[1 + 1{\cdot}15\,\mathrm{e}^{-10{\cdot}2t}\sin(17{\cdot}5t - 120^\circ)] \end{aligned} \qquad (3.3)$$

Here we have treated a complex root in the same manner as a real root; the alternative method is to split the expression into partial fractions, one of which is quadratic:

$$C(s) = \frac{7{\cdot}54}{s} - \frac{7{\cdot}54s + 153{\cdot}8}{s^2 + 20{\cdot}4s + 408}$$

This can be rearranged in the form

$$C(s) = \frac{7{\cdot}54}{s} - \frac{7{\cdot}54s + 153{\cdot}8}{(s + 10{\cdot}2)^2 - 104 + 408}$$

$$= \frac{7{\cdot}54}{s} - \frac{7{\cdot}54(s + 10{\cdot}2)}{(s + 10{\cdot}2)^2 + 304} - \frac{76{\cdot}9}{(s + 10{\cdot}2)^2 + 304}$$

The inverse transform can now be obtained from tables and use of the theorem:

Inverse transform of $F(s + \alpha) = e^{-\alpha t}$ multiplied by the inverse transform of $F(s)$.

The transient responses of eqns. (3.1), (3.2) and (3.3) are shown in Fig. 3.3, where the magnitude of the step function has been adjusted to give the same steady-state output voltage in each case.

Standard Form of Solution

There are many problems having the closed-loop transfer function of the form

$$\frac{C}{R}(s) = \frac{1}{s^2 + Bs + D} \tag{3.4}$$

It is convenient to obtain the solution of this equation for a step input of R when the coefficient B is zero. Then

$$C(s) = \frac{1}{s(s^2 + D)}$$

giving in the time domain,

$$c(t) = 1 - \cos \sqrt{D}\, t$$

This represents a continuous oscillation of angular frequency $\sqrt{D}$ rad/s. This is defined as the *natural frequency* of the system and is denoted by ω_n.

If, in eqn. (3.4), $B^2 = 4D$ the equation reduces to the form

$$\frac{C}{R}(s) = \frac{1}{(s + \sqrt{D})^2}$$

This condition is termed *critical damping*, an instance of which was given in the previous example. The damping ratio is defined as

$$\zeta = \frac{\text{Coefficient of } s}{\text{Coefficient of } s \text{ for critical damping}}$$

For critical damping, $B = 2\sqrt{D}$, or $2\omega_n$.

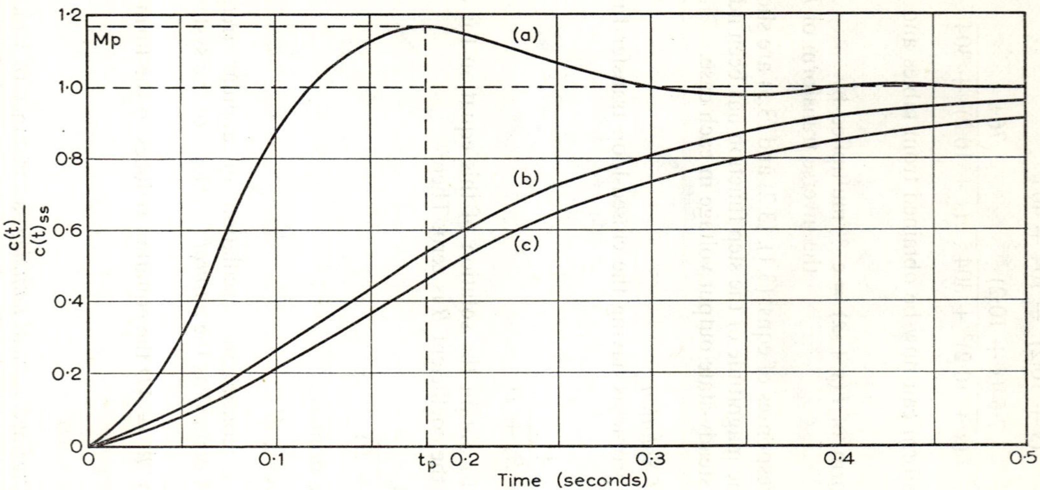

FIG 3.3

(*a*) Open-loop gain constant = 50
(*b*) Open-loop gain constant = 12
(*c*) Open-loop gain constant = 10

$c(t)_{ss}$ = steady-state value of $c(t)$

Hence in general the coefficient of s can be written $2\zeta\omega_n$ and eqn. (3.4) becomes

$$\frac{C}{R}(s) = \frac{1}{s^2 + 2\zeta\omega_n s + \omega_n^{\,2}}$$

In general the numerator of eqn. (3.4) is not unity and this expression becomes

$$\frac{C}{R}(s) = \frac{K\omega_n^{\,2}}{s^2 + 2\zeta\omega_n s + \omega_n^{\,2}} \tag{3.5}$$

where K is a constant.

The solution to this equation for a step-function input signal is

$$c(t) = K\left[1 - \frac{e^{-\zeta\omega_n t}}{\sqrt{(1-\zeta^2)}} \sin\{\omega_n\sqrt{(1-\zeta^2)}t + \cos^{-1}\zeta\}\right] \tag{3.6}$$

or alternatively,

$$c(t) = K\left[1 + \frac{e^{-\zeta\omega_n t}}{\sqrt{(1-\zeta^2)}} \sin\{\omega_n\sqrt{(1-\zeta^2)}t - \psi\}\right]$$

where

$$\psi = \tan^{-1}\frac{\sqrt{(1-\zeta^2)}}{-\zeta}$$

Eqn. (3.6) is plotted for various values of ζ against a dimensionless time parameter $(\omega_n t)$ in standard works on control engineering.

For values of ζ less than unity the response exhibits a damped oscillation of angular frequency

$$\omega_d = \omega_n\sqrt{(1-\zeta^2)} \text{ rad/s} \tag{3.7}$$

The peak value of $c(t)$ is

$$M_p = 1 + e^{-\zeta\pi/\sqrt{(1-\zeta^2)}} \tag{3.8}$$

which occurs at time

$$t_p = \frac{\pi}{\omega_n\sqrt{(1-\zeta^2)}} \quad = \quad \pi/\omega_D \tag{3.9}$$

For part (c) of the preceding example the closed-loop transfer function was

$$\frac{C}{R}(s) = \frac{3080}{s^2 + 20{\cdot}4s + 408}$$

Comparison with the standard form (eqn. (3.5)) yields $\omega_n = 20{\cdot}2$ rad/s and $\zeta = 0{\cdot}5$.

Hence from eqn. (3.7), $\omega_d = 17{\cdot}5$ rad/s, which corresponds to the value given in eqn. (3.3).

Application of eqns. (3.8) and (3.9) gives

$M_p = 1{\cdot}16$ at time $t_p = 0{\cdot}18$ s.

Steady-state Conditions

It is often of interest to know the behaviour of the system following the decay of the initial transient. In the time domain this is given by the particular integral of the differential equation. It is also easily obtained by using the final-value theorem of the Laplace transform:

$$\lim_{t \to \infty} f(t) = \lim_{s \to 0} s\,F(s)$$

where $F(s)$ is the Laplace transform of $f(t)$.

Of special interest is the steady-state value of the error. This is given by

$$\lim_{s \to 0} s\,E(s) = \lim_{s \to 0} \frac{s\,R(s)}{1 + G(s)H(s)} \tag{3.10}$$

This equation is used to define the error constants for various types of input, e.g. step, ramp and parabolic.

For a step-function input the steady-state error is given by

$$e(t)_{ss} = \lim_{s \to 0} \frac{1}{1 + G(s)H(s)} = \frac{1}{1 + \lim_{s \to 0} G(s)H(s)} = \frac{1}{1 + K_0}$$

where K_0 is called the *step* (or *position*) *constant.*

Similarly if the input signal has the form of a ramp,

$$K_1 = \lim_{s \to 0} s\,G(s)H(s)$$

giving a steady-state error of $e(t)_{ss} = 1/K_1$, where K_1 is called the *ramp* (or *velocity*) *constant.*

For a parabolic input the steady-state error is

$$e(t)_{ss} = 2/K_2$$

where K_2 is the *parabolic* (or *acceleration*) *constant* and is defined as

$$K_2 = \lim_{s \to 0} s^2 G(s)H(s)$$

The values of the constants K_0, K_1, K_2, depend upon the form of the open-loop transfer function and in particular on the power of s in the denominator corresponding to the number of pure integrations.

This power defines the type number of the system. For example,

$$\frac{K}{(1 + s\tau_1)(1 + s\tau_2)}$$

is called a *type* 0 *system*, and

$$\frac{K}{s^2(1 + s\tau)}$$

is called a *type* 2 *system*.

Example 3.1

In an electrical servo system used to control a rotatable mass, the moment of inertia of the moving parts is 100 kg-m², the motor torque is 1 600 N-m/rad of misalignment and the damping factor is 0·5. If the control wheel is suddenly turned through an angle of 45°, develop an expression for the subsequent motion of the output shaft. Calculate the error remaining after 2 s. (*Bournemouth*)

Solution

The block diagram of the system is shown in Fig. 3.4, where *B* is the

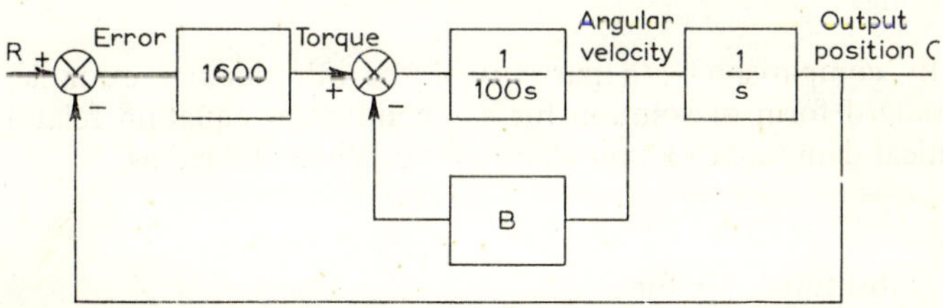

FIG 3.4

effective viscous damping coefficient (N-m/rad-s^{-1}). Simplification of this diagram gives

$$\frac{C}{R} = \frac{1\,600}{100s^2 + Bs + 1\,600}$$

Comparison with the standard form gives $\omega_n = 4$ rad/s. Hence the resulting motion to a step of $\pi/4$ rad is given by

$$c(t) = \frac{\pi}{4}\left[1 - \frac{e^{-\zeta\omega_n t}}{\sqrt{(1 - \zeta^2)}} \sin\{\omega_n\sqrt{(1 - \zeta^2)}t + \cos^{-1}\zeta\}\right]$$

$$= \frac{\pi}{4}\left\{1 - 1{\cdot}15\, e^{-2t} \sin\left(3{\cdot}46t + \frac{\pi}{3}\right)\right\}$$

The output after 2 s is given by substituting $t = 2$ in the above expression. This gives an output angle of $\pi/4$, or 1·016 rad. Hence the error is $45(1{\cdot}016 - 1) = 0{\cdot}74°$.

Example 3.2

A component consists of two non-interacting first-order lags with equal time-constants τ second. Show that if the input is subjected to a step displacement the output will complete 98% of the step in $5{\cdot}83\tau$ second. (*Lanchester*)

Solution

The component has the transfer function $G(s)$, given by

$$G(s) = \frac{1}{(1 + s\tau)^2} = \frac{\dfrac{1}{\tau^2}}{s^2 + \dfrac{2}{\tau} s + \dfrac{1}{\tau^2}}$$

Comparison with the standard form gives

$$\omega_n = \frac{1}{\tau} \qquad \text{and} \qquad \zeta = 1$$

By comparison with part (*b*) of the introductory example, the standard form of solution for a second-order equation relating to critical damping to a unit step input can be obtained as

$$c(t) = 1 - e^{-\omega_n t}(1 + \omega_n t)$$

or, substituting $1/\tau$ for ω_n,

$$c(t) = 1 - e^{-t/\tau}\left[1 + \frac{t}{\tau}\right]$$

Putting $t = 5{\cdot}83\tau$,

$$c(t) = 1 - e^{-5{\cdot}83}(1 + 5{\cdot}83) = 0{\cdot}98$$

Example 3.3

A negative-feedback amplifier consists of three *RC* coupled stages whose midband voltage gains are 20, 15 and 10 respectively. The fraction of the output voltage fed back in antiphase to the input voltage is 1/50. If the r.m.s. input voltage is of 0·2 V magnitude at the midband frequency, calculate the r.m.s. output voltage at this frequency.

A breakdown causes a valve in the first stage to be replaced, which reduces the midband voltage gain of that stage from 20 to 16. Calculate the percentage change in output voltage due to the valve replacement. Derive any formula used and explain briefly the advantages of using feedback in amplifiers. (*Manchester Univ.*)

Solution

The amplifier is represented in block diagram form in Fig. 3.5. The

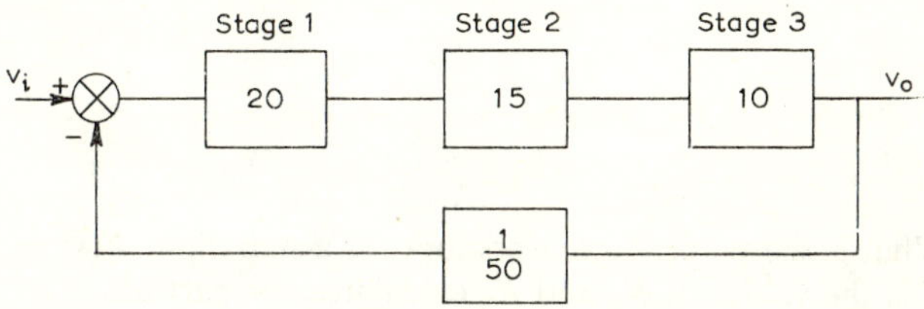

FIG 3.5

voltage gain is given by

$$\frac{v_o}{v_i} = \frac{20 \times 15 \times 10}{1 + \dfrac{20 \times 15 \times 10}{50}} = 49{\cdot}2$$

The output voltage for $v_i = 0{\cdot}2$ is therefore 9·84 V. When the valve in the first stage is replaced, the expression for gain becomes

$$\frac{v_o}{v_i} = \frac{16 \times 15 \times 10}{1 + \dfrac{16 \times 15 \times 10}{50}} = 49{\cdot}0$$

The percentage change in gain and output voltage is therefore

$$\frac{0{\cdot}2}{49{\cdot}2} \times 100 = 0{\cdot}4\%$$

Example 3.4

Fig. 3.6 is a block diagram of a process control system. The process $P(s)$ has the transfer function

$$P(s) = \frac{1}{s(s+1)}$$

The gains K_1 and K_2 control the amount of error and error rate fed to the process.

(*a*) Calculate values of K_1 and K_2 to give a closed-loop transfer function with a second-order response, damping ratio 0·6 and frequency of damped oscillation 10 rad/s.

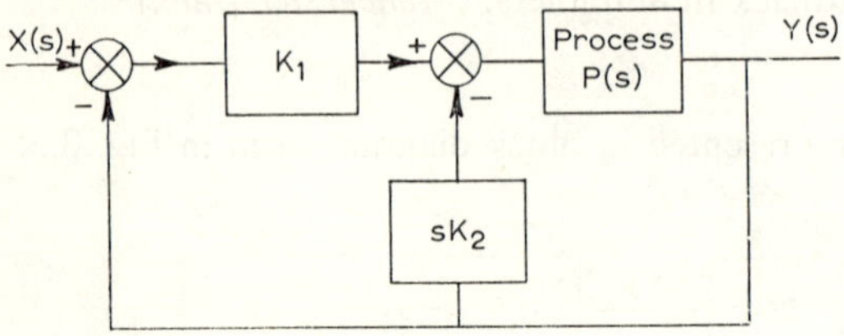

FIG 3.6

(*b*) What is the percentage overshoot for a step input $X(s) = 1/s$? Use the values of K_1 and K_2 calculated for part (*a*).

(*c*) What is the time for first and second crossovers in response to a step input $X(s) = 1/s$? Again use the values of K_1 and K_2 used in (*a*). (*Swansea*)

Solution

The closed-loop transfer function can be obtained by manipulation of the block diagram as shown in Figs. 3.7 and 3.8.

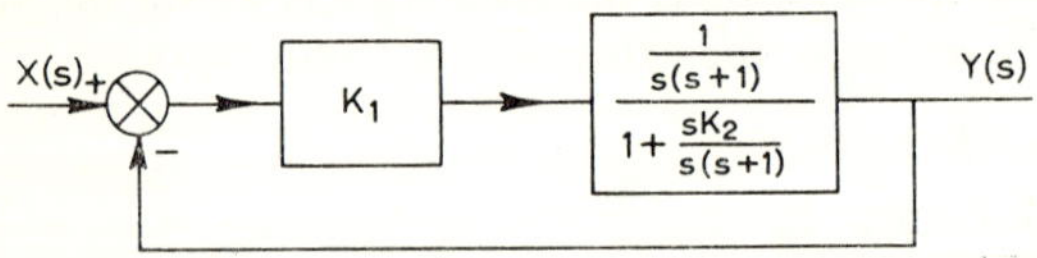

FIG 3.7

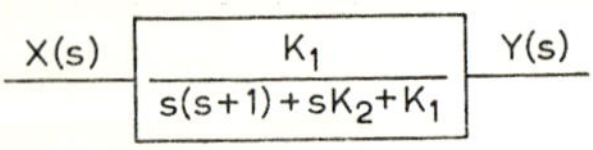

FIG 3.8

(*a*) Hence

$$\frac{Y(s)}{X(s)} = \frac{K_1}{s^2 + (K_2 + 1)s + K_1}$$

Comparing this with the standard form of eqn. (3.5),

$$\omega_n{}^2 = K_1 \qquad 2\zeta\omega_n = K_2 + 1$$

Substituting the value of ω_d in eqn. (3.7),

$$\omega_n = \frac{10}{\sqrt{(1 - 0{\cdot}6^2)}} = 12{\cdot}5\,\text{rad/s}$$

whence

$$K_1 = 156 \qquad \text{and} \qquad K_2 = 14$$

(*b*) The expression for peak overshoot is given by eqn. (3.8). Hence percentage overshoot

$$= e^{-\frac{0\cdot 6\pi}{0\cdot 8}} \times 100 = 9{\cdot}5\%$$

(*c*) The response in the time domain to a step input is given by eqn. (3.6):

$$y(t) = 1 - \frac{e^{-\zeta\omega_n t}}{\sqrt{(1 - \zeta^2)}} \sin\{\omega_n\sqrt{(1 - \zeta^2)}t + \cos^{-1}\zeta\}$$

Equating this expression to unity and solving for t will give the cross-over points, whence

$$t = 0{\cdot}22\,\text{s} \qquad \text{and} \qquad 0{\cdot}53\,\text{s}$$

Example 3.5

The characteristic equation of a second-order dynamic system may be expressed

$$(s^2 + 2\zeta\omega_n s + \omega_n^2)C(s) = 0$$

where $C(s)$ is the Laplace transform of the controlled variable $c(t)$.

(i) Explain the significance of the parameters ζ and ω_n. A servo-mechanism is required to position a load having a polar moment of inertia J and a viscous friction coefficient f. The property of the amplifier-motor combination is that it produces a torque on the load which is K times the angular error, where K is a constant.

(ii) Show that, if the input shaft of the servo-mechanism is rotated at a constant angular velocity ω_i rad/s, the steady-state velocity lag is $2\zeta\omega_i/\omega_n = \omega_i f/K$.

(iii) If $J/f = 0{\cdot}5$ s, determine the value of ω_n such that the velocity lag shall not exceed 1° at an angular speed of 1·26 rad/s.

(iv) Show how the factor ζ could be increased to a value of 0·5 without increasing the velocity lag. Under these conditions, derive an expression for the angular error when the input is of the form $0{\cdot}5\alpha t^2$, where α is a constant acceleration in rad/s^2 and t is time in seconds.

(v) If f is now considered to be zero, find the steady-state acceleration error for $\alpha = 12{\cdot}0\,\text{rad/s}^2$. (*Bangor*)*

Solution

The block diagram is shown in Fig. 3.9, and in a simplified form in

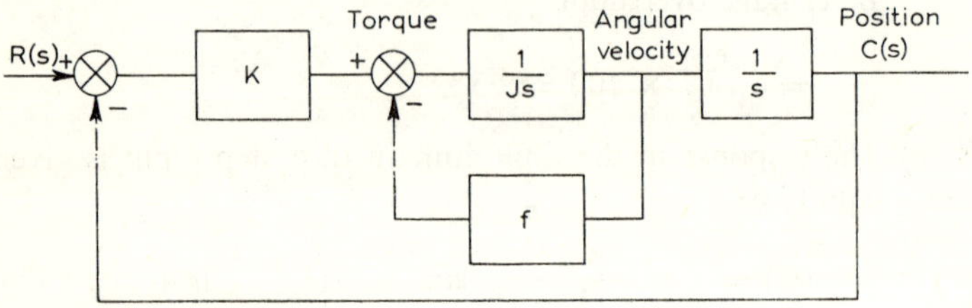

FIG 3.9

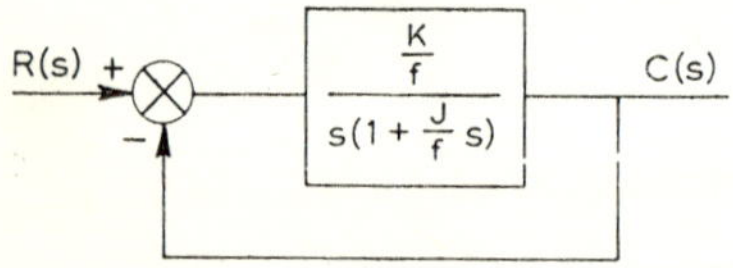

FIG 3.10

Fig. 3.10. The characteristic equation can be found from the closed-loop transfer function

$$\frac{C(s)}{R(s)} = \frac{K/J}{s^2 + sf/J + K/J}$$

Cross-multiplying and putting $R(s) = 0$,

$$(s^2 + sf/J + K/J)C(s) = 0$$

Comparison with the standard form yields

$$f/J = 2\zeta\omega_n \qquad K/J = {\omega_n}^2$$

The steady-state error to a ramp input of position (a continuous rotation of the input shaft at constant angular velocity) is obtained by applying the final-value theorem to the expression for $E(s)$:

$$E(s) = \frac{R(s)}{1 + \dfrac{K/f}{s(1 + (J/f)s)}}$$

Thus in the steady state

$$e(t) = \lim_{s\to 0} s\left[\frac{\omega_i s(s + f/J)}{s^2(s^2 + sf/J + K/J)}\right] = \frac{\omega_i f}{K} \tag{1}$$

$$= \frac{2\zeta\omega_i}{\omega_n}$$

As $J/f = 0{\cdot}5\,\text{s}$, $\zeta\omega_n = 1$.

The steady-state velocity lag can be written $2\omega_i/\omega_n{}^2$. Substituting the appropriate values for ω_i and the lag gives

$$\omega_n = \left(\frac{2 \times 1{\cdot}26 \times 180}{\pi}\right)^{1/2} = 12\,\text{rad/s}$$

It is convenient to draw the block diagram in the form shown in Fig. 3.11. A phase-advance network has also been added to the sys-

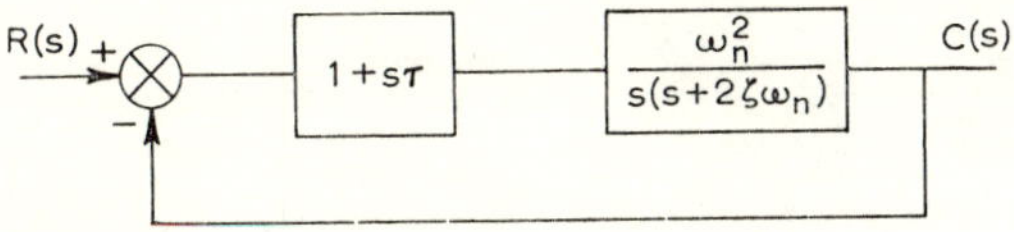

FIG 3.11

tem. Note that this can be approximated by the circuit shown in Fig. 1.1 (*c*).

It can be shown by following the above procedure that the characteristic equation is

$$[s^2 + (2\zeta\omega_n + \omega_n{}^2\tau)s + \omega_n{}^2] = 0$$

The effective damping ratio is now $\zeta + \omega_n\tau/2$. Given that $\omega_n =$ 12 rad/s, that $\zeta = 1/12$ and that the effective damping ratio is to be 0·5, the required value of τ is 0·07s.

The error-input relation is

$$\frac{E(s)}{R(s)} = \frac{s^2 + 2\zeta\omega_n s}{s^2 + (2\zeta\omega_n + \omega_n{}^2\tau)s + \omega_n{}^2}$$

and comparison with eqn. (1) reveals that the velocity lag is unchanged.

When the input has the form $0{\cdot}5\alpha t^2$,

$$R(s) = \frac{\alpha}{s^3}$$

and

$$E(s) = \frac{\alpha(s + 2\zeta\omega_n)}{s^2[s^2 + (2\zeta\omega_n + \omega_n{}^2\tau)s + \omega_n{}^2]} \tag{2}$$

As the above system is type 1 it cannot follow a parabolic input; this would be shown if the final value theorem had been applied to eqn. (2). However, if the feedback damping term is zero the block diagram of Fig. 3.11 exhibits two integrators (a characteristic of a type 2 system) and the steady-state acceleration error is

$$\lim_{s\to 0} s\left[\frac{s\alpha}{s^2(s^2+\omega_n^2\tau s+\omega_n^2)}\right] = \frac{\alpha}{\omega_n^2} = \frac{1}{12}$$

Example 3.6

The diagram in Fig. 3.12 shows part of a wind-tunnel aircraft pitch-control system. The pitch angle is θ_o and the pilots input is θ_i. V_v is the vertical velocity signal and θ_e the elevator angle.

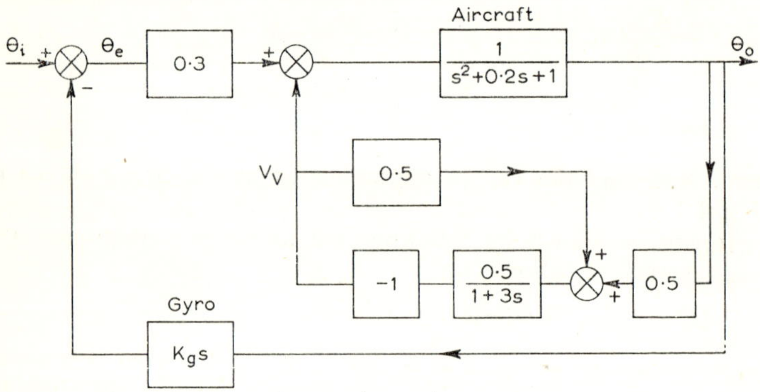

FIG 3.12

By simplifying the diagram, or otherwise, determine the closed-loop transfer function θ_o/θ_i and thence, or otherwise, from the resulting error transfer function θ_e/θ_i, determine the steady-state error to a constant velocity input at θ_i of 0·1 rad/s. (*Brunel Univ.*)

Solution

The successive steps in the reduction of the block diagram are shown in Figs. 3.13 and 3.14. Reduction of the block diagram in Fig. 3.14 gives the overall transfer function

$$\frac{\theta_o}{\theta_i} = \frac{0{\cdot}3(1{\cdot}25+3s)}{3s^2+(1{\cdot}85+0{\cdot}9K_g)s^2+(3{\cdot}25+0{\cdot}375K_g)s+1{\cdot}5}$$

The error/input relationship is

$$\frac{\theta_e}{\theta_i} = \frac{1}{1 + \dfrac{0{\cdot}3(1{\cdot}25 + 3s)K_g s}{3s^3 + 1{\cdot}85s^2 + 3{\cdot}25s + 1{\cdot}5}}$$

$$= \frac{3s^3 + 1{\cdot}85s^2 + 3{\cdot}25s + 1{\cdot}5}{3s^3 + (1{\cdot}85 + 0{\cdot}9K_g)s^2 + (3{\cdot}25 + 0{\cdot}375K_g)s + 1{\cdot}5}$$

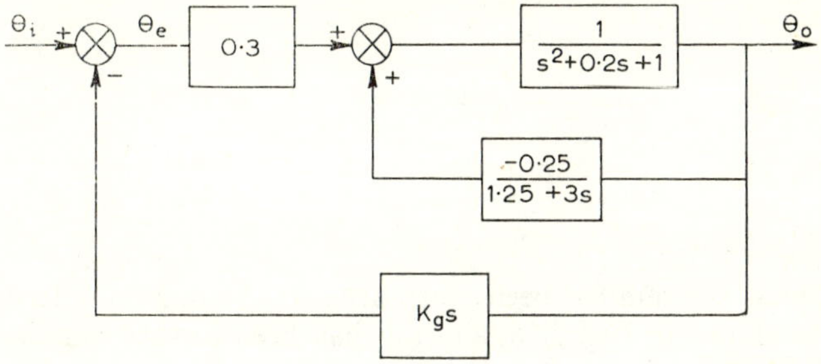

FIG 3.13

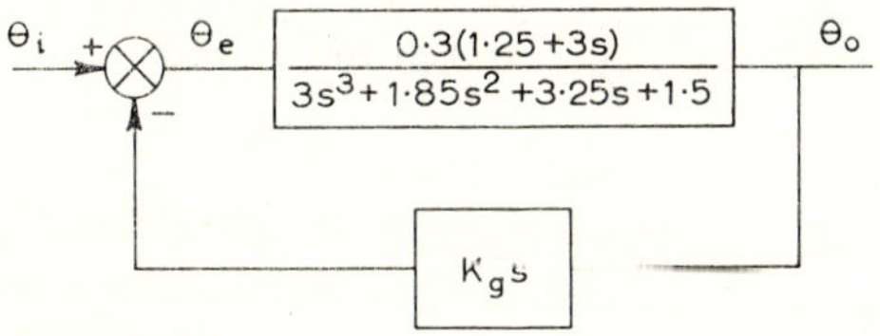

FIG 3.14

Because of the $K_g s$ factor in the feedback path, θ_i controls the output velocity.

The steady-state value of θ_e is given by

$$\theta_e = \lim_{s \to 0} s \frac{0{\cdot}1}{s} \frac{3s^3 + 1{\cdot}85s^2 + 3{\cdot}25s + 1{\cdot}5}{3s^3 + (1{\cdot}85 + 0{\cdot}9K_g)s^2 + (3{\cdot}25 + 0{\cdot}375K_g)s + 1{\cdot}5}$$

$$= 0{\cdot}1 \text{ rad}$$

Example 3.7

A servomechanism can be represented by the block diagram shown in Fig. 3.15. Prove that the steady-state error to a ramp input of magnitude A is $A(\tau - \tau_3 + 1/K)$.

With the numerical data $K = 30$, $\tau_1 = 0{\cdot}01$ s, $\tau_2 = 0{\cdot}02$ s, $\tau_3 = 0{\cdot}005$ s and $\tau = 0{\cdot}04$ s, and given that the two real roots are at

$s = -17{\cdot}8$ and $s = -240$, determine the closed-loop transfer function with a factorized denominator. (*Nottingham Univ.*)

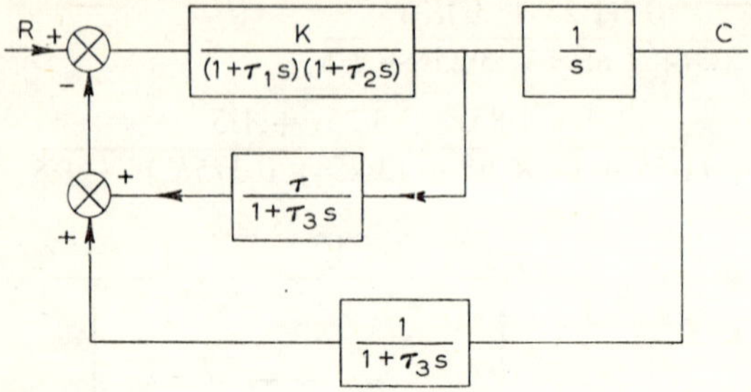

FIG 3.15

Solution

The block diagram has been rearranged to obtain a unity feedback path as shown in Fig. 3.16, and this has been further simplified in Fig. 3.17. The relation

$$\frac{C}{E}(s) = \frac{K(1 + s\tau_3)}{s[(1 + s\tau_1)(1 + s\tau_2)(1 + s\tau_3) + K(\tau - \tau_3)]}$$

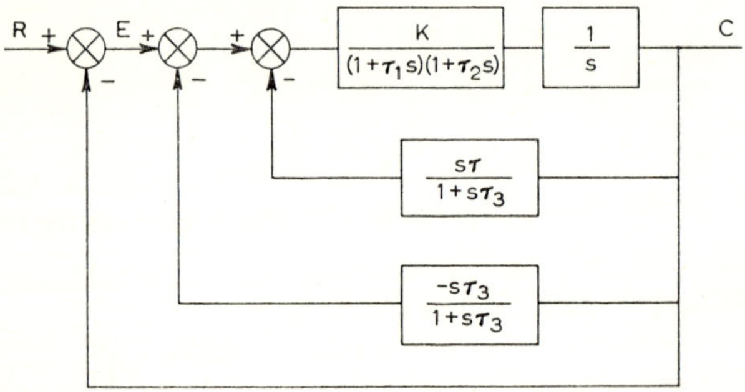

FIG 3.16

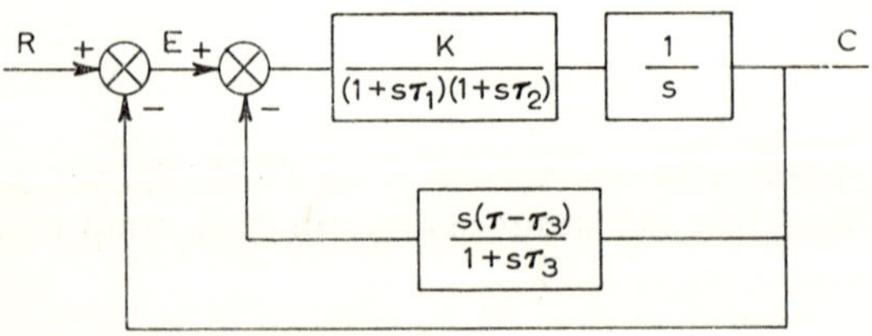

FIG 3.17

is easily obtained by reduction of the inner loop. Hence

$$\frac{E}{R}(s) = \frac{1}{1 + C/E(s)}$$

$$= \frac{s[(1 + s\tau_1)(1 + s\tau_2)(1 + s\tau_3) + K(\tau - \tau_3)]}{s[(1 + s\tau_1)(1 + s\tau_2)(1 + s\tau_3) + K(\tau - \tau_3)] + K(1 + s\tau_3)}$$

For a ramp input of magnitude A, $R(s) = A/s^2$ and the steady-state error is

$$e(t)_{ss} = \lim_{s \to 0} s\left(\frac{A}{s^2}\right)\left[\frac{s[(1 + s\tau_1)(1 + s\tau_2)(1 + s\tau_3) + K(\tau - \tau_3)]}{s[(1 + s\tau_1)(1 + s\tau_2)(1 + s\tau_3) + K(\tau - \tau_3)] + K(1 + s\tau_3)}\right]$$

$$= A[1/K + \tau - \tau_3]$$

The closed-loop transfer function is given by

$$\frac{C}{R}(s) = \frac{\dfrac{K(1 + s\tau_3)}{s(1 + s\tau_1)(1 + s\tau_2)(1 + s\tau_3) + Ks(\tau - \tau_3)}}{1 + \dfrac{K(1 + s\tau_3)}{s(1 + s\tau_1)(1 + s\tau_2)(1 + s\tau_3) + Ks(\tau - \tau_3)}}$$

$$= \frac{K(1 + s\tau_3)}{s(1 + s\tau_1)(1 + s\tau_2)(1 + s\tau_3) + Ks\tau + K}$$

Inserting the numerical values of the various quantities and simplifying,

$$\frac{C}{R}(s) = \frac{30(1 + 0{\cdot}005s)}{s^4(10^{-6}) + s^3(3{\cdot}5 \times 10^{-4}) + s^2(3{\cdot}5 \times 10^{-2}) + s(2{\cdot}2) + 30}$$

In factorized form the closed-loop transfer function is

$$\frac{C}{R}(s) = \frac{30(1 + 0{\cdot}005s) \times 10^6}{(s + 17{\cdot}8)(s + 240)(s + 46 + j70)(s + 46 - j70)}$$

Example 3.8

Regulation of the output voltage of a constant-speed d.c. shunt generator is to be effected using a closed-loop system in which a fraction K_F of the terminal voltage is compared with a very stable 85 V reference input. The difference is amplified by an amplifier having a gain of K_A and its output is connected to the generator field.

Calculate the values of K_A and K_F in order that the steady-state terminal voltage with a load of 20 A shall be 0·1 V less than the no-load value of 250 V.

If, whilst the regulated generator is running unloaded, a resistive load of 12·5 Ω is suddenly supplied, determine the time taken for the terminal voltage to recover to 245 V.

The generator constants are as follows:

Field resistance, 200 Ω
Field inductance, 40 H
Armature resistance, 0·7 Ω
Armature inductance, negligible.

A steady current of 1 A in the field produces a no-load terminal voltage of 250 V. (*Brunel Univ.*)

Solution

The block diagram of the system is shown in Fig. 3.18. As the system is linear we can determine the effects of the load current and the input voltage on the generated voltage separately.

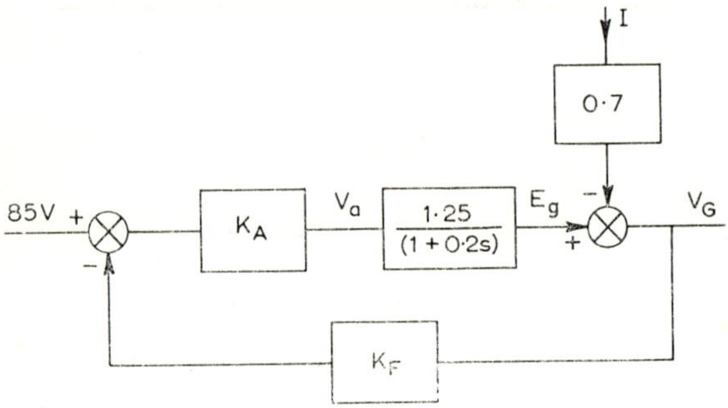

FIG 3.18

To obtain the relation between the change of terminal voltage V_g due to I we redraw the block diagram as in Fig. 3.19, where $m = K_A K_F(1{\cdot}25)$. Thus

$$\frac{V_g}{(0{\cdot}7)I} = -\frac{(1 + 0{\cdot}2s)}{(1 + m)\left(1 + \dfrac{0{\cdot}2}{1 + m}s\right)}$$

the steady-state condition being

$$\frac{V_g}{I} = -\frac{0{\cdot}7}{1 + m}$$

Substitution of the values $I = 20$ A, $V_g = -0{\cdot}1$ V yields $m = 139$.

The second steady-state relation available is that, between the steady-state output voltage and the input reference on no-load,

$$\frac{250}{85} = \frac{K_A(1{\cdot}25)}{1+m}$$

Thus $K_A = 329$ and $K_F = 0{\cdot}338$. The complete relation is

$$\frac{V_g}{I} = -\frac{0{\cdot}005(1+0{\cdot}2s)}{1+(0{\cdot}2/140)s}$$

If the load resistance suddenly changes by 12·5 Ω the effect is to apply a step function of load current of magnitude 20 A. (Note that

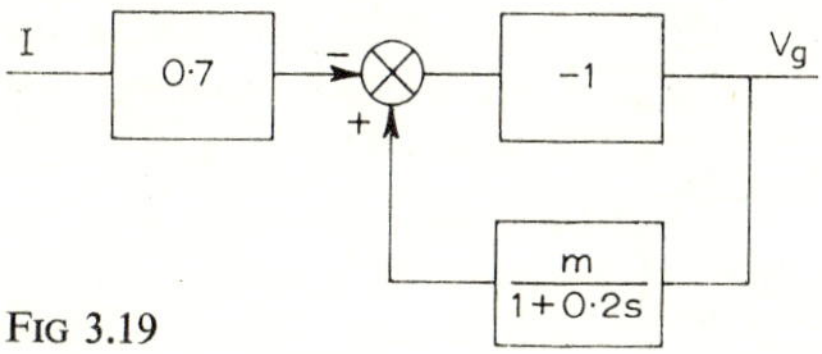

FIG 3.19

this step can be regarded as a constant 20 A after the load is applied, the change in V_g being small compared with its absolute value.) The corresponding change in the generator voltage V_g is

$$\frac{-0{\cdot}1(1+0{\cdot}2s)}{s\left(1+\dfrac{0{\cdot}2s}{140}\right)} = -\frac{14(s+5)}{s(s+700)}$$

$$= -\frac{0{\cdot}1}{s} - \frac{13{\cdot}9}{s+700}$$

Taking inverse Laplace transforms,

$$v_g(t) = -0{\cdot}1 - 13{\cdot}9\,\mathrm{e}^{-700t}$$

Note that at $t = 0$ the instantaneous fall of voltage is 14 V, which is equal to 20(0·7), the iR drop in the machine.

The time taken for the voltage to recover from a fall of 5 V is given by

$$\mathrm{e}^{-700t} = 4{\cdot}9/13{\cdot}9 \qquad \text{i.e.} \qquad t = 1{\cdot}5\,\text{ms}$$

Example 3.9

Important parameters in a control system are stability, accuracy and speed of response. Briefly discuss the compromises a designer must make.

Draw the block diagram of the control system shown in Fig. 3.20. Determine the loop transfer function and from this find the steady-state velocity error for a step-velocity disturbance of 1·0 rad/s to the input of the system.

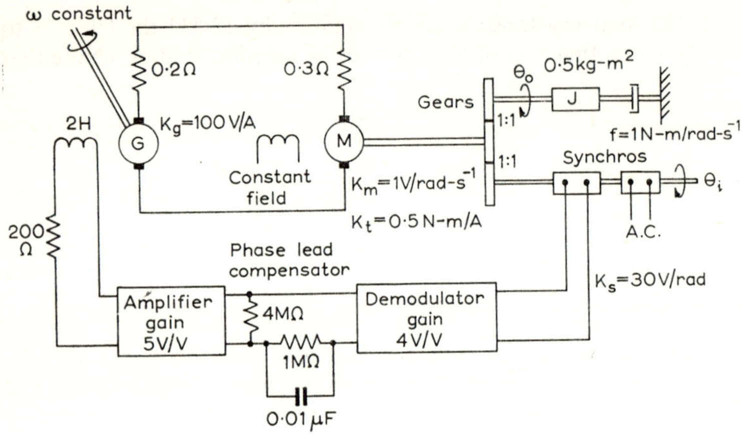

FIG 3.20

To obtain critical damping, what alteration must be made to the phase-lead compensator without changing the value of the velocity-error constant, and what would be the value of the capacitor which would result in critical stability? (*Swansea*)*

Solution

Generally both the accuracy and the speed of response are improved by increasing the open-loop gain of the system. This leads to instability, and care must be taken to ensure that methods of stabilization used do not mar the performance of the system.

The block diagram is shown in Fig. 3.21. The open-loop transfer function is

$$\frac{\theta_0}{E}(s) = \frac{120}{s(1 + 0{\cdot}008s)(1 + 0{\cdot}25s)}$$

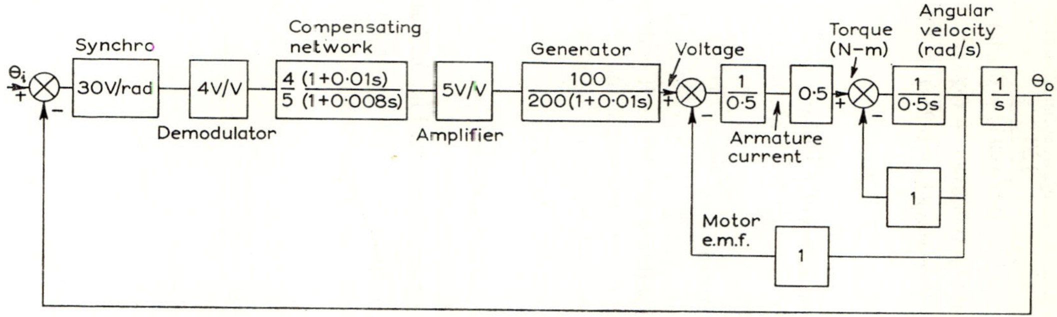
θi
+
−
Synchro
30V/rad
4V/V
Demodulator
Compensating network
4/5 (1+0·01s)/(1+0·008s)
5V/V
Amplifier
Generator
100/200(1+0·01s)
Voltage
+
−
1/0·5
Armature current
0·5
Torque (N−m)
+
−
Angular velocity (rad/s)
1/0·5s
1
1/s
θo
Motor e.m.f.
1

Fig 3.21

The steady-state velocity error for a step-velocity disturbance is given by

$$e(t)_{ss} = \lim_{s\to 0} s\frac{1}{s^2}\frac{1}{1+\dfrac{120}{s(1+0{\cdot}008s)(1+0{\cdot}25s)}}$$

$$= \lim_{s\to 0}\frac{1}{s}\left(\frac{s(1+0{\cdot}008s)(1+0{\cdot}25s)}{s(1+0{\cdot}008s)(1+0{\cdot}25s)+120}\right)$$

$$= \frac{1}{120}\,\text{rad}$$

where $\theta_i = 1/s^2$ for a unit ramp of position.

Alternatively the error is given by $1/K_v$, where K_v is the velocity constant and is given by

$$\lim_{s\to 0} s\,G(s)H(s) = 120\,\text{rad}^{-1}$$

The question of critical stability will be dealt with in Chapter 4.

By altering the value of the capacitor the amount of damping in the system will be altered without changing the velocity-error constant.

Example 3.10

The parameters of the system shown in Fig. 3.22 are:

Tacho-generator constant, $K_1 = 1{\cdot}0\,\text{V/rad-s}^{-1}$
Current feedback constant, $K_2 = 0{\cdot}001\,\text{V/A}$
Amplifier gain, $K_3 = 90$
Amplifier time-constant, $\tau_A = 0{\cdot}057\,\text{s}$

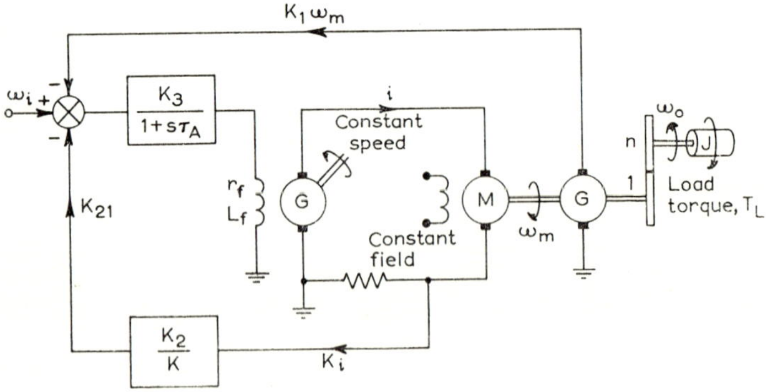

FIG 3.22

Generator field circuit, $r_f = 40\,\Omega$, $L_f = 0{\cdot}75$ H
Generator e.m.f. constant; $K_4 = 100$ V/A
Motor e.m.f. constant, $K_5 = 4$ V/rad-s^{-1}
Generator-motor armature circuit, $r_a = 0{\cdot}04\,\Omega$, L_a negligible
Motor torque constant, $K_6 = 3{\cdot}8$ N-m/A
Total inertia referred to load shaft, $J = 1\,355$ kg-m^2
Gear ratio, $n = 2{\cdot}5/1$
Frictional torque, negligible.

Derive the general expression relating the output velocity ω_o to the load torque T_L and input signal ω_i. Give the corresponding steady-state relationship when ω_i and T_L are constant.

Explain briefly why, when the load torque is small, the system controls ω_o rather than i. What influence does the current-feedback constant K_2 have upon the magnitude of the current i and the speed regulation of the system? What analogy can be drawn between the current feedback in this system and velocity feedback in a position-control system?

If the input signal ω_i is 50 V d.c. find (i) the steady-state no-load value of ω_o, (ii) the steady-state value of the standstill torque.

What are the gain and phase margins of this system?

Suggest some non-linear form of current feedback which might allow good speed regulation and yet give adequate protection against current overload. (*Sheffield Univ.*)*

Solution

The expression relating the output velocity ω_o, load torque T_L and input signal ω_i is obtained by applying Mason's rule to the signal-flow graph (see Chapter 2).

The block diagram is shown in Fig. 3.23 and the associated signal-flow graph in Fig. 3.24, where

$$G_1 = \frac{90}{(1 + 0{\cdot}057s)} \qquad G_2 = \frac{100}{40(1 + 0{\cdot}19s)}$$

$$G_3 = \frac{1}{0{\cdot}04} \qquad G_4 = 3{\cdot}8$$

$$G_5 = \frac{1}{1\,355s} \qquad G_6 = \frac{1}{2{\cdot}5}$$

$$G_7 = -\frac{1}{2{\cdot}5}$$

$$H_1 = -0{\cdot}001 \qquad H_2 = -4$$

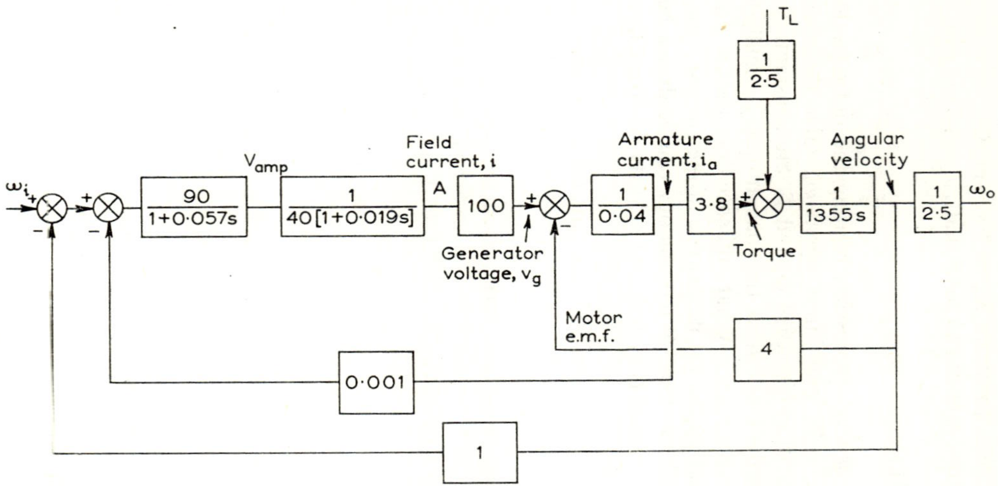

ω_i
+
−
90
1+0·057s
V_{amp}
1
40[1+0·019s]
Field
current, i
A
100
Generator
voltage, v_g
1
0·04
Armature
current, i_a
3·8
Torque
T_L
1
2·5
1
1355s
Angular
velocity
1
2·5
ω_o
Motor
e.m.f.
4
0·001
1

FIG 3.23

The relation between the quantities is found by the following steps. The graph determinant is

$$\Delta = 1 - [(-G_1G_2G_3G_4G_5) + G_1G_2G_3H_1 + G_3G_4G_5H_2]$$

The forward-path transmittance between ω_i and ω_o is

$$G_1G_2G_3G_4G_5G_6 \qquad \text{and} \qquad \Delta_1 = 1 - (0)$$

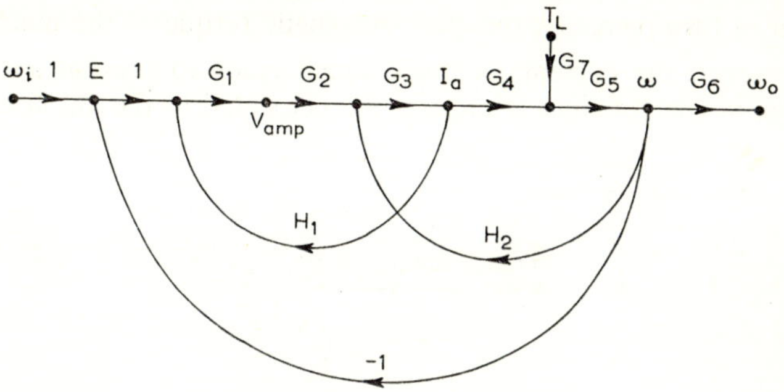

FIG 3.24

since there are no closed paths when the above forward-transmittance path is removed. Thus

$$\frac{\omega_o}{\omega_i} = \frac{G_1G_2G_3G_4G_5G_6}{1 + G_1G_2G_3G_4G_5 - G_1G_2G_3H_1 - G_3G_4G_5H_2}$$

The relation between T_L and ω_o is found by determining the new forward-path transmittance $G_5G_6G_7$ and the co-factor

$$\Delta_1 = 1 - (G_1G_2G_3H_1)$$

Thus

$$\frac{\omega_o}{T_L} = \frac{G_5G_6G_7(1 - G_1G_2G_3H_1)}{1 + G_1G_2G_3G_4G_5 - G_1G_2G_3H_1 - G_3G_4G_5H_2}$$

Substituting the values, using superposition, and simplifying gives

$$\omega_o = \frac{22{\cdot}5\omega_i - 2{\cdot}36 \times 10^{-3}[1 + 0{\cdot}18(1 + 0{\cdot}057s)(1 + 0{\cdot}19s)]T_L}{(1 + 0{\cdot}057s)(1 + 0{\cdot}19s)(1 + 3{\cdot}57s) + 56(1 + 0{\cdot}36s)} \qquad (1)$$

The steady-state relationship is obtained by putting $s = 0$:

$$\omega_o = \frac{22{\cdot}5\omega_i - 2{\cdot}36 \times 10^{-3}(1 + 0{\cdot}18)T_L}{1 + 56}$$

$$= 0{\cdot}393\omega_i - 4{\cdot}9 \times 10^{-5}T_L \qquad (2)$$

For small load torques it can be seen from the block diagram of Fig. 3.25 that the current feedback quantity will be smaller than the velocity feedback.

With zero load torque the motor current is also zero and ω_i is sufficient to cancel ω and provide the necessary error signal to the amplifier.

For $T_L > 0$, ω_o decreases and the resulting increased error in signal in turn increases the electromagnetic torque of the machine.

However, the current feedback signal tends to prevent a rapid change in current since it reduces the net signal to the amplifier. If

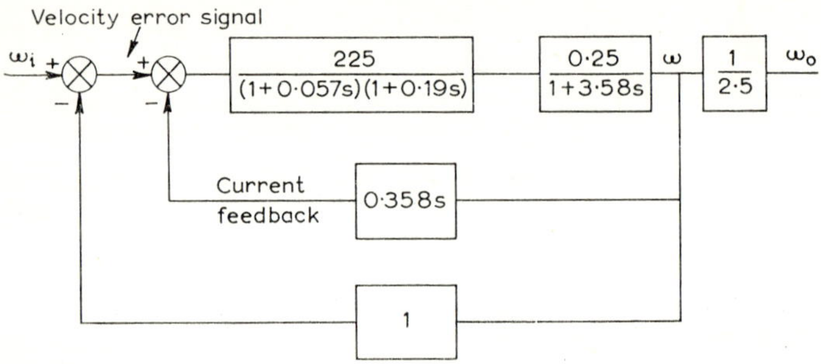

FIG 3.25

K_2 were sufficiently large the motor could be stalled without damage since the current signal would cancel ω_i, limiting the current to say 2·5 times the full-load value. Unfortunately, as has been mentioned above, the curre:.t build-up will be reduced by this action and the motor could not respond quickly to suddenly applied loads. Insertion of a dead zone in the current feedback path, if suitably arranged, will only let a signal pass when the current has reached a certain value, say twice full-load current. Under normal conditions the current feedback path will be open.

Inspection of Fig. 3.25 shows that the current feedback in the velocity control system is a derivative feedback quantity. This is analogous to velocity feedback in a position control system.

Substituting $\omega_i = 50\,\text{V}$ and $T_L = 0$ in eqn. (2) gives the no-load speed as $\omega_o = 19{\cdot}6\,\text{rad/s}$.

The stalled torque can be obtained by putting $\omega_o = 0$ in this equation, whence $T_L = 400\,000$-N-m.

For gain and phase margins see Chapter 6.

Example 3.11

(*a*) Give a definition of sensitivity. Hence determine expressions for the sensitivity of the transmittance T of the system shown in Fig. 3.26 to changes in G and H. Comment on the significance of the expressions.

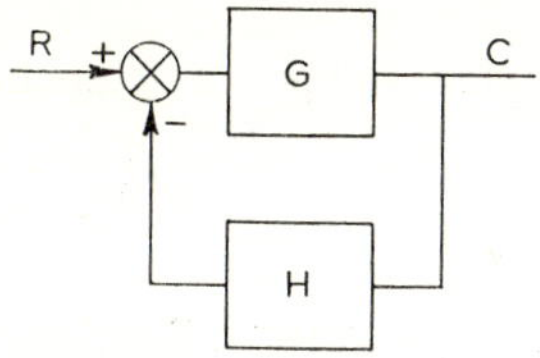

FIG 3.26

(*b*) In a position control system the forward path transfer function is $100/s(1 + s)$ and the feedback-path transfer function is 10. Determine the sensitivity of T with respect to the feed-forward and feedback elements respectively in the vicinity of $\omega = 1$ rad/s. (*Lanchester*)

Solution

(*a*) The *sensitivity* of the system may be defined as

$$\frac{\text{Change in system transmittance } T}{\text{Change in open-loop parameter which causes change in } T}$$

Here change is defined as the ratio of the differential variation of the function to the actual function itself. An ideal system has zero sensitivity being independent of changes in the elements of the system. Here the sensitivity of T with respect to G is

$$S_G{}^T = \frac{dT}{T}\Big/\frac{dG}{G} = \frac{G}{T}\left[\frac{1}{(1 + GH)^2}\right] = \frac{1}{1 + GH} \tag{1}$$

The sensitivity of T with respect to H is

$$S_H{}^T = \frac{dT}{T}\Big/\frac{dH}{H} = \frac{H}{T}\left[\frac{-G^2}{(1 + GH)^2}\right] = \frac{-GH}{1 + GH} \tag{2}$$

As the open-loop function $GH \gg 1$, then $S_G{}^T \to 0$, $S_H{}^T \to -1$. The results show that (i) the higher the open-loop gain the less is the sensitivity of the system transmittance to change in the forward path; (ii) although GH must be large it need not be precise or stable; (iii) the elements in the feedback path must be precise and stable since changes in their characteristics are reflected in the system transmittance.

The final point justified the designer's rule that a system variable cannot be controlled more accurately than it can be measured.

(*b*) In this case, $G = 100/s(1 + s)$, $H = 1$, and from eqn. (1),

$$S_G{}^T = \frac{s(s+1)}{s(s+1)+100}$$

Substituting $j\omega$ for s, we obtain the frequency-dependent function

$$S_G{}^T(j\omega) = \frac{j\omega(1+j\omega)}{j\omega(1+j\omega)+100}$$

the magnitude of which at $\omega = 1$ rad/s is $1{\cdot}4 \times 10^{-3}$. Similarly

$$S_H{}^T(j\omega) = \frac{-100}{j\omega(1+j\omega)+100} \approx -1$$

PROBLEMS

3.1 A servo system for the positional control of a rotatable mass is stabilized by viscous-friction damping which is three-quarters of that needed for critical damping. The undamped frequency of the system is 12 Hz. Derive an expression for the output of the system if the input control is suddenly moved to a new position, the system being initially at rest. Hence find the amount of the first overshoot. (*Bournemouth*)

Ans. $\theta_o(t) = \theta_i[1 - 1{\cdot}5\,e^{-56{\cdot}6t}\sin(50t + 0{\cdot}72)]$; 2·8%.

3.2 Draw a block diagram of the components of a simple electrical servomechanism for controlling the position of a rotatable mass. Assume that viscous-friction damping is used and develop the characteristic differential equation. Explain the operation of the system, the need for damping, and by reference to the forms of transient response, why critical damping is desirable.

In such a system the moment of inertia of the moving parts is 0·4 kg-m^2 and the motor torque is 1 000 N-m per radian of misalignment. The friction torque is 30 N-m/rad-s^{-1}. Determine by how much the friction torque must be increased in order that the damping should be critical. (*Bournemouth*)

Ans. 10 N-m/rad-s^{-1}.

3.3 In the position-control system shown in Fig. 3.27, damping is by viscous friction. Explain briefly how the system operates and develop the characteristic differential equation.

If the amplifier voltage gain is 150, the error-detecting device produces 2·5 V per degree of error, the inertia referred to the output shaft is 0·08 kg-m^2, the motor torque is 0·04 N-m per 100 V applied, find the frictional torque at the output shaft required to produce a

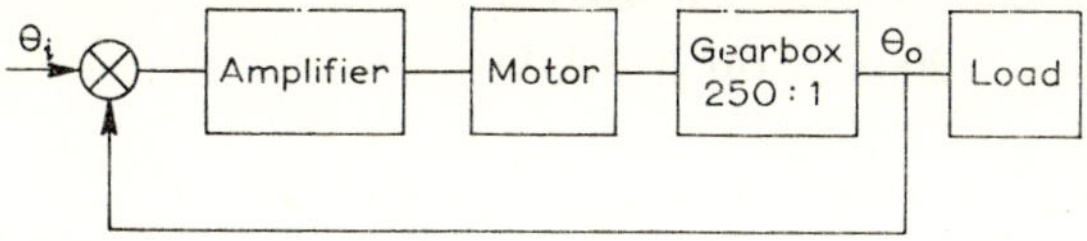

FIG 3.27

damping factor of 0·5 (i.e. half critical damping). For a constant input speed of 15·7 rad/s, calculate the steady angular lag of the output shaft. (*Bournemouth*)*

Ans. 13·2 N-m/rad-s^{-1}; 5·5°.

3.4 A servomechanism, designed to control the angular position of a rotable mass, is stabilized by means of acceleration feedback. The moment of inertia of the system is 10^{-5} kg-m^2, the viscous frictional torque per radian per second is 10^{-4} N-m and the motor torque T_m is given by

$$T_m = 4 \times 10^{-3}\left(e + k\frac{d^2\theta_o}{dt^2}\right) \text{ newton-metres}$$

where $e = (\theta_i - \theta_o)$ is the angular positional error in radians between input and output shafts, and $k\,d^2\theta_o/dt^2$ is the additional feedback signal.

(i) Draw the block diagram of the system.
(ii) Establish the differential equation for the system.
(iii) Determine the value to be assigned to k in order that the damping shall be critical.
(iv) Determine the value of the steady-state error for an input signal which has a constant angular velocity of 1·26 rad/s.

Derive any formula used. (*I.E.R.E.*)*

Ans. (iii) 2·34 × 10^{-3} s^2; (iv) 0·0314 rad.

3.5 The block diagram (Fig. 3.28) shows a servo system designed to follow a constant-speed input shaft. The synchro-control transformer has an error voltage gradient of 1 V/deg, and the speed of the d.c. permanent-magnet servomotor is approximately proportional to armature current at a rate of 314 rad-s^{-1}/A.

(*a*) Determine the velocity coefficient (k_v) for a velocity error of 0·1° when the shaft speed is 2·1 rad/s.

(*b*) What value of amplifier gain will produce the required value of velocity coefficient?

(*c*) Show how a d.c. tachogenerator may be used to improve the system performance, or describe, with diagrams, a suitable phase-sensitive detector. (*I.E.R.E.*)*

Ans. (*a*) $k_v = 1\,200\,\mathrm{s}^{-1}$; (*b*) 20/3 A/V.

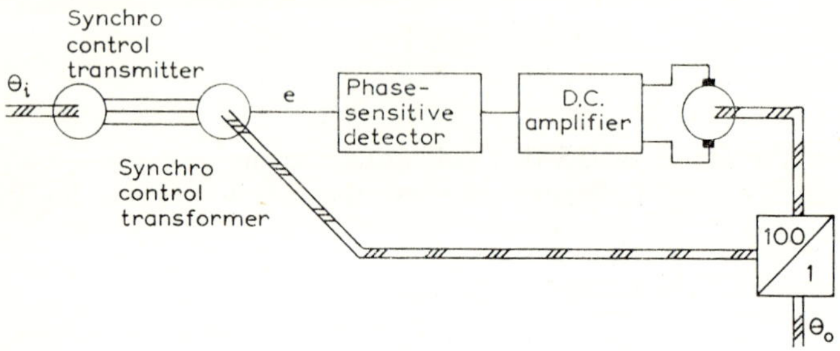

FIG 3.28

3.6 The output of a simple position-control servomechanism has both inertia and viscous drag, the latter being proportional to the velocity of the output. If the controller is actuated by the error signal between the input and output signals, and if the system is linear, use the method of the Laplace transform to obtain the response of the system, from rest, to a unit-step change of input position control.

Deduce a relationship between the constants of the system such that the "overshoot" of the output in response to the unit step is equal to 10% of the final value of the output step. (*I.E.R.E.*)

Ans. $f = 1{\cdot}2\sqrt{(KJ)}$.

3.7 A Ward Leonard system is used for controlling the speed of a d.c. motor. The field of the motor to be controlled is supplied at constant voltage, and the voltage applied to the armature of the motor is derived from a d.c. generator. This generator is driven at constant speed, and has its field supplied from a d.c. electronic amplifier. The input signal to this amplifier is equal to the difference between a steady reference voltage V_{in} and a voltage V_{out} generated by a tachogenerator which is coupled to the motor shaft. The details of the system are as follows:

Generator. Resistance of the field circuit, including the output resistance of the d.c. amplifier, is 200 Ω; generated e.m.f. per ampere of field current is 2 000 V.

Amplifier. Voltage gain on open-circuit is 20; input resistance is to be assumed infinite.

Motor. Torque developed per ampere of armature current is 1·0 N-m; generated e.m.f. per radian per second is 1·0 V; armature resistance, including resistance of generator armature, is 2·0 Ω.
Tachogenerator. Generated e.m.f. per radian per second of motor speed is 0·5 V.

(i) Draw a simplified circuit diagram for the system.
(ii) Derive the equation relating the motor speed ω to the reference voltage V_{in} and the torque T.
(iii) What is the speed when $V_{in} = 101$ V and T is zero?
(iv) By what percentage of the no-load speed does the speed fall when $T = 10$ N-m, assuming $V_{in} = 101$ V?
(v) What is the effect on the no-load speed should the gain of the electronic amplifier be reduced by 20%, assuming $V_{in} =$ 101 V? (*I.E.R.E.*)

Ans. (ii) $\omega = \frac{200}{101} V_{in} - \frac{2}{101} T$; (iii) 200 rad/s; (iv) 0·099%; (v) speed is reduced to 199·4 rad/s.

3.8 A velodyne speed control system consists of a d.c. electronic amplifier which supplies the field winding of a field controlled d.c. motor. The armature current of this motor is kept constant. Coupled to the motor shaft is a tachogenerator which produces a voltage V_o. The d.c. amplifier receives at its input a voltage $V_i - V_o$ in which V_i is a d.c. reference voltage derived from a potential divider connected across a constant-voltage d.c. source. The parameters of the system are as follows:

Electronic amplifier: voltage gain on open-circuit is 1000; output resistance is 2 kΩ; input resistance is infinite.
D.C. motor: resistance of field winding is 2 kΩ; total motor torque T is 10 N-m per ampere of field current.
Tachogenerator: sensitivity 2·5 V per radian per second of motor speed.

(i) Draw a simplified circuit diagram for the system.
(ii) Derive an equation relating the motor speed ω_o to the reference voltage V_i and the total motor torque T.
(iii) If the reference voltage is 100 V, calculate the no-load speed and the percentage regulation when the motor is producing a total torque of 1 N-m.
(iv) By what percentage does the speed change if the resistance of the field winding of the motor increases by 20% when the reference is fixed at 100 V and the motor is providing a fixed total torque of 1 N-m? (*I.E.R.E.*)

Ans. (ii) $\omega = 0{\cdot}4V_i - 0{\cdot}16T$; (iii) 0·4%, (iv) 0·0402%

3.9 A schematic diagram of a simple position-control system is shown in Fig. 3.29. It may be assumed that the only dynamically significant energy storage is associated with the moments of inertia of motor, gearing and load, and that the only viscous damping torque developed at the motor shaft arises from the action of the tachogenerator. The output resistance of the potentiometers [voltage

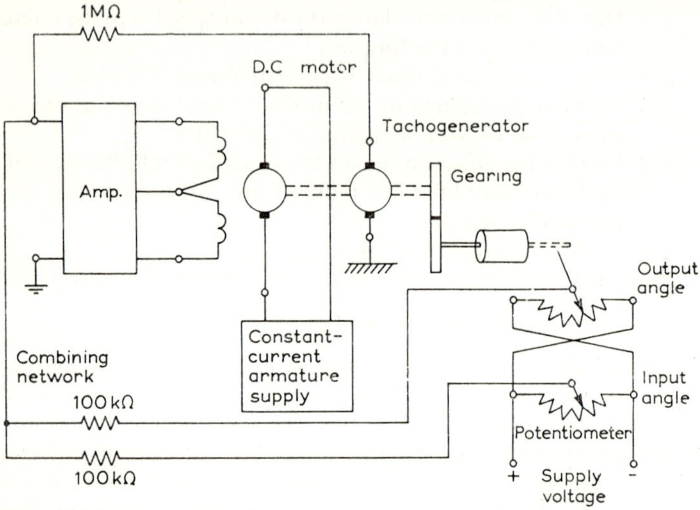

FIG 3.29

dividers] may be neglected in comparison with the resistances of the resistive combining network. For the set of system parameters given below derive:

(*a*) the differential equation describing output shaft motion, and
(*b*) the response of the output shaft to a step change of one radian in input angle position, with the system initially at rest.

The system parameters are:

Maximum angle of travel of potentiometers, 6 rad.
Voltage across setting and resetting potentiometers, 20 V.
Moment of inertia of load about output shaft, 0·5 kg-m^2.
Total moment of inertia of motor, tachogenerator and gearing referred to motor shaft, 0·005 kg-m^2.
Gearbox step-down ratio, 10 to 1.
Tachogenerator constant, 2 V per radian per second.
Amplifier transfer conductance, 0·5 A/V.
Motor torque constant, 1·26 N-m/A. (*London Univ.*)

Ans. (*a*) $\dfrac{10}{s^2 + 6s + 10}$; (*b*) $c(t) = 1 - 0{\cdot}316\, e^{-3t} \sin (t + 0{\cdot}31)$.

3.10 (*a*) Fig. 3.30 shows two generators cascaded together. The appropriate constants are also given. Deduce a transfer function for the output/input voltage ratio. Deduce the output signal when a 10 V step is applied to the first generator.

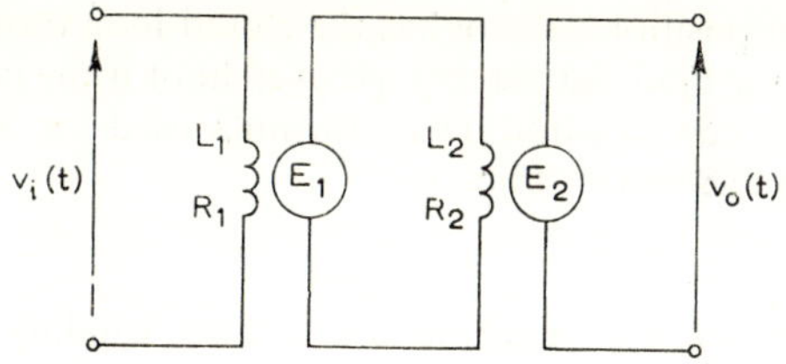

FIG 3.30

$L_1 = 100$ H $\quad L_2 = 50$ H
$R_1 = 200\ \Omega$ $\quad R_2 = 25\ \Omega$
$K_1 = 250$ V/A (field) $\quad K_2 = 100$ V/A (field)
Generator back-e.m.f. to be taken as zero

(*b*) Fig. 3.31 shows a system with the same generators. Deduce the value of $\zeta = R_2/(R_1 + R_2)$ to produce a critically damped response, when $K = 1{\cdot}0$.

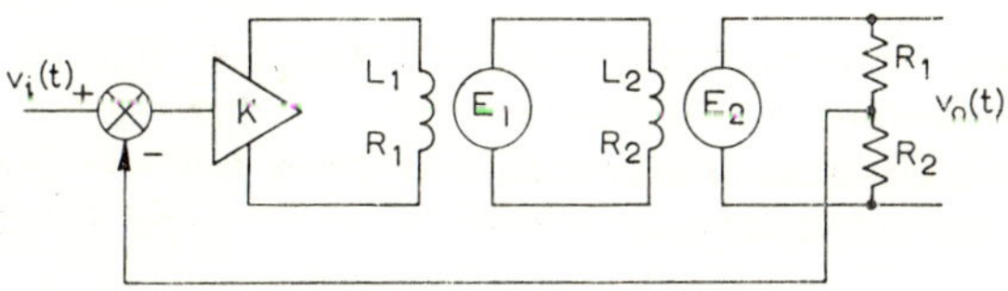

FIG 3.31

(Assume there is no current drain on the output generator due to resistor chain $R_1 R_2$.) (*Leicester*)

Ans. (*a*) $\dfrac{5}{(1 + 0{\cdot}5s)(1 + 2s)}$; $v_o(t) = 50[1 + 0{\cdot}33\ e^{-2t} - 1{\cdot}33\ e^{-0{\cdot}5t}]$

(*b*) $\zeta = 0{\cdot}11$.

3.11 The angular position of a rotatable mass, driven by an electric motor, is controlled from an input potentiometer [voltage divider] using a closed-loop automatic control system. Draw a block diagram and set up the differential equation of the system assuming there is viscous friction.

Sketch waveforms showing the output response to be expected from a sudden change in input when the system is (i) overdamped, (ii) critically damped, (iii) underdamped.

If the moment of inertia of the moving parts is 900 kg-m² and the viscous friction 1 900 N-m per rad/s, what must be the overall gain measured as torque per radian of misalignment in order for the system to be critically damped? (*Leicester*)

Ans. $K = 1\,000$ N-m/rad.

3.12 A servo system for position control has the closed-loop transfer function $6/(s^2 + 2s + 6)$. Find the percentage overshoot if the input is suddenly moved to a new position. (Any formula used for overshoot must be justified.) (*Scottish Assoc.*)

Ans. 24·4%.

3.13 Fig. 3.32 represents a d.c. machine whose field winding has inductance L and resistance R. The moment of inertia at the shaft

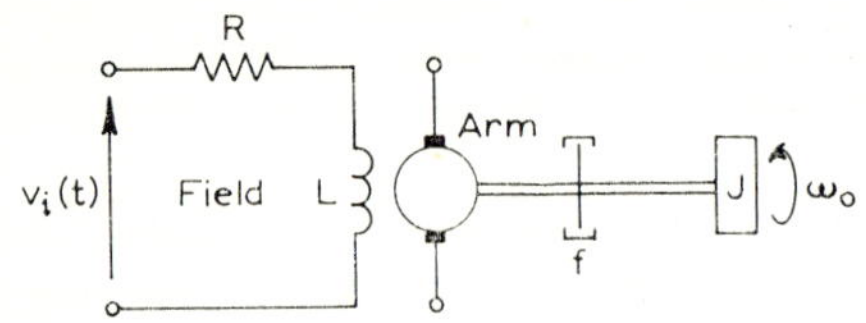

FIG 3.32

is J and the viscous friction coefficient is f. Assuming that the flux is directly proportional to the field current and that the motor is provided with a constant armature current, show that the Laplace transform of the output angular velocity ω_o to the input voltage v_i is given by

$$\frac{\omega_0}{v_i}(s) = \frac{K}{JL(s + R/L)(s + f/J)}$$

where K is a constant.

If $R = 50\,\Omega$, $L = 5$ H, $J = 0{\cdot}5$ kg-m², $f = 0{\cdot}5$ N-m/rad-s^{-1} and $K = 100$ N-m per field ampere, find an expression relating the instantaneous angular velocity of the load to a suddenly applied input voltage of 5 V. Sketch the graph of ω_o against time. (*Scottish Assoc.*)

Ans. $\omega_o(t) = 20(1 - 1{\cdot}11\,e^{-t} + 0{\cdot}11\,e^{-10t})$

3.14 Fig. 3.33 is a block diagram of a simple remote position servomechanism which drives a load through a gearbox of ratio n:1. The moment of inertia of the moving parts is J; f is the viscous friction coefficient; and KK_e is the motor torque per radian of misalignment. Establish the differential equation of the system when it is stabilized by velocity feedback, the tachogenerator having a constant of K_v volts per radian per second.

In this control system the moment of inertia is 4 000 kg-m², the motor torque is 800 N-m/V and the error constant is 1 V/rad. If the gearbox ratio is 20:1 and the tachogenerator constant is 0·03 V per rad/s calculate, for critical damping, (i) the natural angular frequency, (ii) the coefficient of viscous friction, and (iii) the steady-state velocity error for an input of 1 rad/s. (*Scottish Assoc.*)

Ans. (i) 2 rad/s; (ii) 16 N-m/rad-s^{-1}, (iii) 1 rad.

3.15 A separately excited d.c. motor is supplied by a constant-voltage 600 V supply. The armature circuit resistance is 0·1 Ω and and the associated inductance is 0·005 H. The moment of inertia

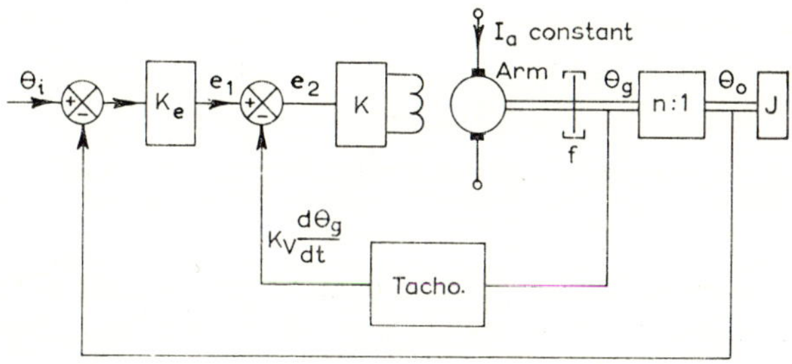

FIG 3.33

referred to the motor shaft is 80 kg-m². The speed of the motor is adjusted to 52·4 rad/s. Draw a transfer-function block diagram of the system relating motor torque (N-m) to speed (rad/s).

If the load torque is suddenly increased by 2 500 N-m determine the following:

(i) the undamped natural angular frequency (rad/s)
(ii) the system damping ratio
(iii) the frequency of the transient speed oscillation (Hz)
(iv) the steady-state drop in motor speed (rad/s). (*Lanchester*)*

Ans. (i) 18 rad/s; (ii) 0·55; (iii) 2·4 Hz, (iv) 1·9 rad/s.

3.16 Fig. 3.34 shows a control system with output acceleration feedback which is needed to meet the system specification. Show, however, that if $N(s) = 1$ and there is negligible viscous friction, the system is unstable, and find the natural angular frequency if $K = 10$, $K_m = 2$ and $J = 4$ in appropriate units.

Show that if $N(s)$ is of the form $(A + s\tau)/s$, the system may be stabilized whilst preserving the acceleration feedback. (*Brunel Univ.*)

Ans. 0·91 rad/s.

3.17 Compare the effects of phase advance compensation and the use of derivative of output feedback on the steady-state error to a ramp input signal for an ideal type 1 system. (*Lanchester*)

Ans. Uncompensated system: $E/R = \dfrac{1}{1 + K/s}$, $e(t)_{ss} = 1/K$

Compensated with $(1 + \tau_1 s)/(1 + \tau_2 s)$, $e(t)_{ss} = 1/K$
With feedback compensation, $K_2 s$, $e(t)_{ss} = 1/K + K_2$

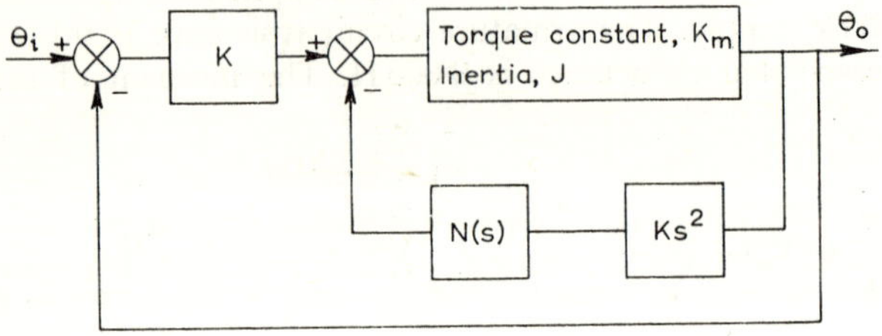

FIG 3.34

3.18 Two otherwise similar position control systems have different forms of stabilization. The first, having velocity feedback, has the characteristic equation

$$\left(\frac{s^2}{\omega_0^2} + \tau s + 1\right)\theta_0 = \theta_i$$

The second incorporates a phase-advance network and has the equation

$$\left(\frac{s^2}{\omega_0^2} + \tau s + 1\right)\theta_0 = (1 + \tau s)\theta_i$$

where in both cases $\omega_0/2\pi$ is the undamped natural frequency, τ is the damping time-constant, θ_i is the input angular position and θ_o is the output angular position.

Sketch graphs showing the response of these two systems when $\tau\omega_o = 2$ for the following two cases:

(*a*) The input θ_i changes by a step of magnitude H.
(*b*) The input θ_i remains zero but the output θ_o is released suddenly from a displaced position $\theta_o = -H$.

Determine for both systems the steady-state velocity lag when the demanded position is $\theta_i = \Omega t$, where $\Omega = \omega_o/2$ and $\tau\omega_o = 1$. (*I.E.E.*)

Ans. $\frac{1}{2}$, 0.

3.19 Fig. 3.35 shows a Ward Leonard speed-control system using two identical machines. Each machine has two identical field windings of resistance 50 Ω and inductance 50 H, an armature resistance $R_a = 0{\cdot}09\,\Omega$ and is rated at 200 V, 10 A.

The generator (G) is driven at a constant speed of 100 rad/s such that a current of 1 A in either field winding gives an open-circuit output voltage of 100 V. The motor (M) field windings are connected

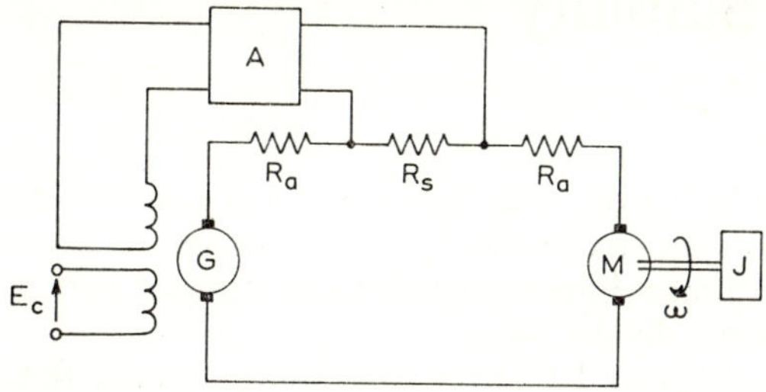

FIG 3.35

in series and supplied with a constant current of 1 A; the motor drives a load J of inertia 2 kg-m².

The voltage developed across the small series resistance R_s of 0·02 Ω is amplified by a factor A and applied to one of the generator field windings so as to oppose large increases in armature current by reducing the generator excitation. The system input, E_c volts, is applied to the other field winding.

If $A = 0$ (i.e. no feedback) show that

$$\frac{I_a}{E_c} = \frac{10s}{(s+1)(s+10)} \qquad \frac{\omega}{I_a} = \frac{1}{s}$$

where I_a = armature current in amperes; ω = motor speed in radians per second.

Draw the block diagram for the system with $A = 445$.

Show that the loop characteristic equation is

$$s^2 + 100s + 10 = 0$$

and that

$$\frac{I_a}{E_c} = \frac{10s}{(s+100)(s+0{\cdot}1)}$$

Hence, if a step input E_c of 100 V is applied with the motor at rest, show that I_a will not exceed the rated current.

Assume throughout that the machines behave linearly and that effects due to armature inductances and load friction can be ignored. (*Nottingham Univ.*)

Chapter 4

Absolute Stability

So far, we have looked at the time response of a system obtained by the solution of a differential equation. In the design of a control system it is of the utmost importance that the transient part of the solution should eventually die away.

We have seen in Chapter 1 that the closed-loop transfer function has the form

$$\frac{C}{R}(s) = \frac{N(s)}{D(s)}$$

where $N(s)$ and $D(s)$ are polynomials in s. Factorizing gives

$$\frac{C}{R}(s) = K\frac{(s - Z_1)(s - Z_2)\ldots(s - Z_n)}{(s - P_1)(s - P_2)\ldots(s - P_n)}$$

When $s = Z_1, Z_2, \ldots Z_n$, $C/R(s)$ is zero; when $s = P_1, P_2, \ldots P_n$, the function becomes infinite.

The former values of s are known as *zeros*, and the latter values of s as *poles*, of the function.

Consider

$$\frac{C}{R}(s) = \frac{(s + 1)(s + 2)}{s(s^2 + 2s + 10)} = \frac{(s + 1)(s + 2)}{s(s + 1 + j3)(s + 1 - j3)}$$

The zeros of this function are at $s = -1$, $s = -2$, and the poles are at $s = 0$, $s = -1 + j3$, $s = -1 - j3$. These are plotted on the s plane in Fig. 4.1.

For any value of s, say s_1, $(C/R)(s)$ has magnitude and argument:

$$\left|\frac{C}{R}(s_1)\right| = \frac{|s_1 + 1||s_1 + 2|}{|s_1||s_1 + 1 + j3||s_1 + 1 - j3|}$$

$$\underline{/\frac{C}{R}(s_1)} = \underline{/s_1 + 1} + \underline{/s_1 + 2}$$

$$- \underline{/s_1} - \underline{/s_1 + 1 + j3} - \underline{/s_1 + 1 - j3}$$

But $|s_1 + 1|$ is the length of the line joining the point s_1 to the zero at -1 and $\underline{/s_1 + 1}$ is the angle made by this line and the real axis. Similarly, the other terms represent lengths and angles associated with lines from the point s_1 to the appropriate poles and zeros. This means that, knowing the positions of the poles and zeros, the magnitude and angle of the function can be obtained graphically for any point on the s plane.

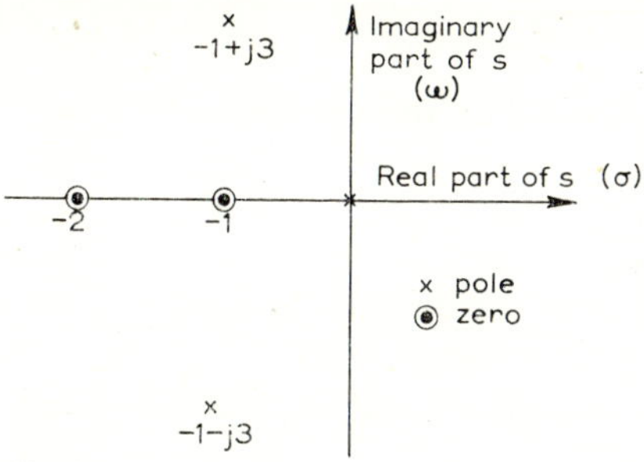

FIG 4.1

In order to obtain the time response $c(t)$ for a given input $r(t)$ the inverse Laplace transform of $C(s)$ is required. This involves splitting $C(s)$ into partial fractions thus:

$$C(s) = \frac{K_1}{(s - P_1)} + \frac{K_2}{(s - P_2)} + \cdots \frac{K_m}{(s - P_m)}$$
$$+ \text{ additional terms due to } R(s)$$

Each term $K_j/(s - P_j)$ represents a term of the form $K_j e^{P_j t}$ in the time response, and in order for this not to increase indefinitely with time, the real part of P_j must be negative. Hence for stability none of the poles must be in the right-hand half of the s plane.

The forcing function $R(s)$ does not affect the position of the system poles and therefore the system stability. For investigation of the absolute stability, any form of $R(s)$ can be assumed, and it is most convenient to take $R(s) = 0$. In this form the differential equation is characteristic only of the system and is known as the *characteristic equation.*

The Nyquist Criterion

This is a graphical method of determining the stability of a system. The closed-loop transfer function is

$$\frac{C}{R}(s) = \frac{G(s)}{1 + G(s)H(s)}$$

where $G(s)H(s)$ is the open-loop transfer function. For stability we have seen that there must be no poles of $\frac{C}{R}(s)$ in the right-hand half of the s plane. This condition is met if the function $B(s) = 1 + G(s)H(s)$ has no zeros having positive real parts, the zeros being the roots of the characteristic equation. Writing

$$B(s) = \frac{(s - Z_1) \ldots (s - Z_n)}{(s - P_1) \ldots (s - P_m)}$$

the poles and zeros can be plotted on the s plane as in Fig. 4.2.

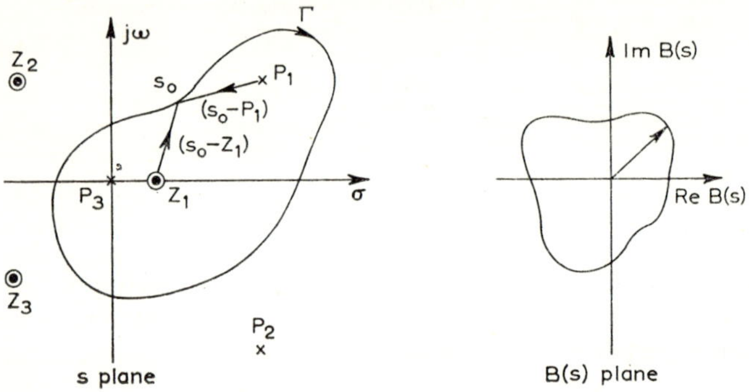

FIG 4.2

Consider a point s_0 on this plane. The value of $B(s)$ at this point is

$$B(s_0) = \frac{(s_0 - Z_1)(s_0 - Z_2) \ldots (s_0 - Z_n)}{(s_0 - P_1)(s_0 - P_2) \ldots (s_0 - P_m)}$$

Each of the terms corresponds to a vector on the s plane. If s_0 takes a succession of values given by the path Γ, the values of $B(s)$ map out a contour on the $B(s)$ plane. The vectors $(s_0 - Z_1)$, $(s_0 - P_1)$, $(s_0 - P_3)$ undergo a rotation of 360° as this path is traversed, while the vectors from any points outside the contour undergo zero rotation. Hence the function $B(s)$ undergoes a net rotation clockwise of $2\pi(Z - P)$ (for anticlockwise rotation considered as positive the number of rotations, N, is $P - Z$), where Z and P are the numbers of zeros and poles, respectively, enclosed by the path Γ. The minus sign is necessary because the poles are due to factors in the denominator of $B(s)$. (Remember that a complex number of the form $(s_0 - Z_1)$ can be represented in polar form by $A\,e^{j\theta}$, and that $1/(s_0 - P_1)$ has the polar form $A\,e^{-j\theta}$.)

As an illustration let us consider the following example. The characteristic equation of a certain system is given by

$$B(s) = \frac{10(s+1)}{s^2 - 4s + 13}$$

Factorizing the denominator,

$$B(s) = \frac{10(s+1)}{(s-2+j3)(s-2-j3)}$$

Hence $B(s)$ has a zero at $s = -1$ and poles at $s = 2 \pm j3$.

Any contour not actually passing through a pole or zero can be chosen to illustrate the theorem, and for selected values of s_0 on the

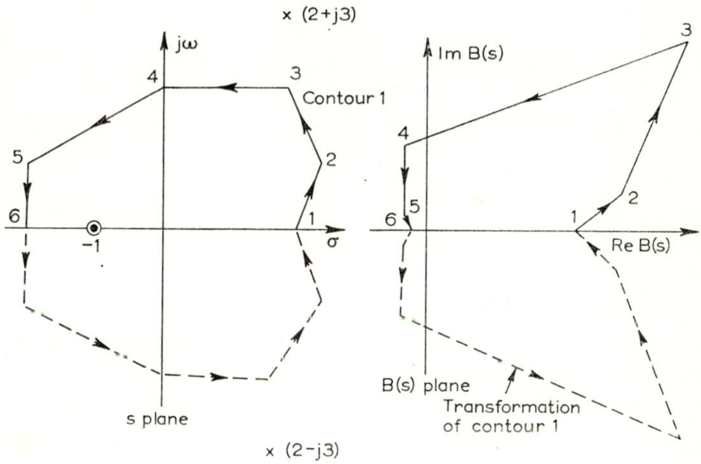

FIG 4.3

contour the appropriate values of $B(s)$ can be calculated. This can be done in two ways, and both are shown taking the point on the contour $s = 2 + j2$ as an illustration.

(i) The value $s = 2 + j2$ can be substituted directly in the formula for $B(s)$ giving

$$B(s) = \frac{10(2 + j2 + 1)}{(j5)(-j1)} = 6 + j4$$

(ii) $B(s)$ can be obtained graphically by measurement of the angles and lengths as explained earlier:

$$|B(s)| = \frac{10 \times 3{\cdot}6}{1 \times 5} = 7{\cdot}2$$

$$\underline{/B(s)} = 34^\circ - (-90^\circ) - 90^\circ = 34^\circ$$

Values of s for two contours on the s plane and the corresponding values of $B(s)$ are given in the following table and are plotted in Figs. 4.3 and 4.4.

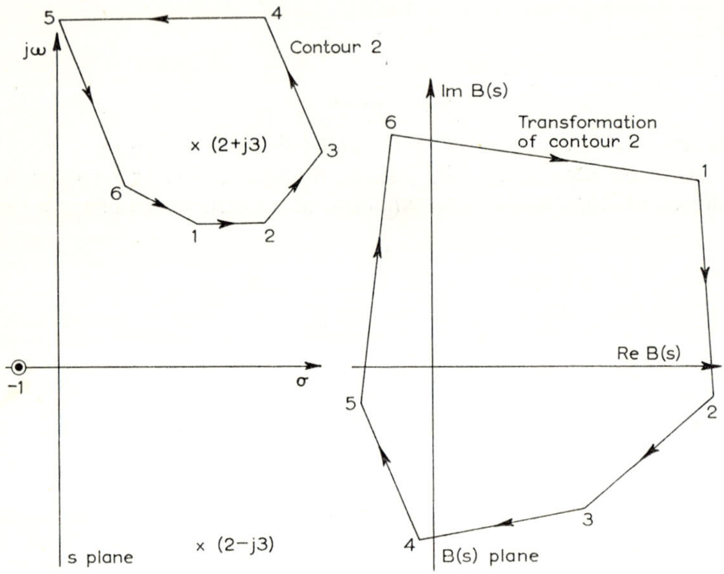

FIG 4.4

Point	Contour 1		Contour 2	
	s	$B(s)$	s	$B(s)$
1	$2+j0$	$3{\cdot}33+j0$	$2+j2$	$6+j4$
2	$2{\cdot}5+j1$	$4{\cdot}35+j0{\cdot}7$	$3+j2$	$6{\cdot}15-j0{\cdot}77$
3	$2+j2$	$6+j4$	$4+j3$	$3{\cdot}5-j3$
4	$0+j2$	$-0{\cdot}48+j1{\cdot}8$	$3+j5$	$-0{\cdot}31-j3{\cdot}5$
5	$-2+j1$	$-0{\cdot}5+j0{\cdot}25$	$0+j5$	$-2{\cdot}06-j0{\cdot}73$
6	$-2+j0$	$-0{\cdot}4+j0$	$1+j2{\cdot}5$	$-1{\cdot}28+j4{\cdot}9$

Half of contour 1 is obtained by using values of s that are complex conjugates of those given in the table, giving a plot of $B(s)$ that is symmetrical about the real axis.

If the points on the contour are taken in order it is seen that:

(*a*) Contour 1 on the *s* plane encircles a zero in an anticlockwise direction, and the corresponding contour on the $B(s)$ plane encircles the origin in an anticlockwise direction.

(*b*) Contour 2 on the *s* plane encircles a pole in an anticlockwise direction, and the corresponding contour on the $B(s)$ plane encircles the origin in a clockwise direction.

If a path is chosen to enclose the right-hand half-plane completely, then any net rotation of $B(s)$ about the origin indicates $Z - P$. As the number of poles of $B(s)$ in the right-hand half-plane can be obtained by inspection or by the use of Routh's rule (to be dealt

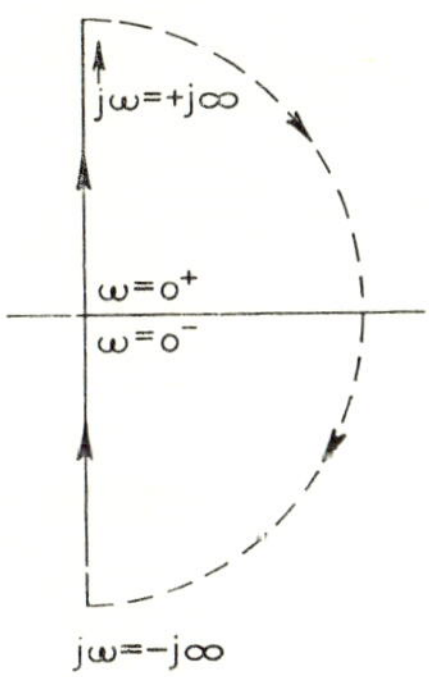

FIG 4.5

with later) the number of zeros in the right-hand half-plane can be determined. Many systems have no poles in the right-hand half-plane, and stability is indicated if $B(s)$ does not enclose the origin. Instead of plotting $B(s) = 1 + G(s)H(s)$, it is easier to plot $G(s)H(s)$ (the open-loop transfer function) and this should not enclose the -1 point for stability.

The particular path chosen is shown in Fig. 4.5. It includes the whole of the imaginary axis and an infinite semicircle enclosing the right-hand half-plane. (The contour must be modified slightly if a pole or a zero lie on the imaginary axis.) It is proved in the standard works that all the rotation of $B(s)$ occurs while the contours traverse the complete imaginary axis, i.e. from $-j\infty$ to $+j\infty$, provided that $\lim_{s \to 0} G(s)H(s) = 0$ or a constant.

Routh's Stability Criterion

This is an algebraic approach to the determination of the absolute stability of a system. Consider a system having a characteristic equation of the form

$$a_n s^n + a_{n-1} s^{n-1} + \ldots + a_0 = 0$$

For stability the following conditions must be satisfied:

(*a*) All coefficients of s from n to zero must be present and positive.
(*b*) There must be no changes of sign in the first column of the array formed from the coefficients of the above equation. Any change of sign indicates a zero of the right-hand half-plane.

$$\begin{array}{l|llll}
s^n & a_n & a_{n-2} & a_{n-4} & \ldots \\
s^{n-1} & a_{n-1} & a_{n-3} & a_{n-5} & \ldots \\
s^{n-2} & b_1 & b_2 & b_3 & \ldots \\
s^{n-3} & c_1 & c_2 & c_3 & \ldots \\
 & \cdot & \cdot & \cdot & \\
 & \cdot & \cdot & \cdot & \\
 & \cdot & \cdot & \cdot & \\
s^1 & l & & & \\
s^0 & m & & &
\end{array}$$

where $b_1 = \dfrac{a_{n-1}a_{n-2} - a_n a_{n-3}}{a_{n-1}}$

$$b_2 = \frac{a_{n-1}a_{n-4} - a_n a_{n-5}}{a_{n-1}}$$

$$c_1 = \frac{b_1 a_{n-3} - a_{n-1} b_2}{b_1}$$

The above table of coefficients is known as a *Routhian array*.

Example 4.1

Describe what is meant by the Nyquist stability criterion. Draw Nyquist contours of the following open-loop transfer functions and discuss briefly their relative stability in the closed loop.

(*a*) $\dfrac{K}{(1 + s\tau_1)(1 + s\tau_2)}$ (*b*) $\dfrac{K}{s(1 + s\tau_1)(1 + s\tau_2)}$

(*c*) $\dfrac{K}{s^2(1 + s\tau_1)}$ (*d*) $\dfrac{K(1 + s\tau_1)(1 + s\tau_2)}{s^3(1 + s\tau_3)(1 + s\tau_4)}$

(*Manchester Univ.*)

Solution

(*a*) $G(s) = K/(1 + s\tau_1)(1 + s\tau_2)$

The form of this type of plot is shown in Fig. 4.6 for two values of K. In sketching this function the following information is available. As

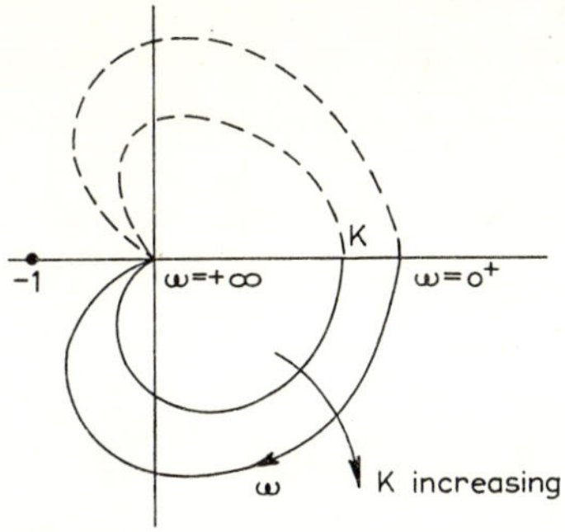

FIG 4.6

the locus in the s plane follows the imaginary axis $s = j\omega$ the function becomes

$$G(j\omega) = \frac{K}{(1 + j\omega\tau_1)(1 + j\omega\tau_2)}$$

For $\omega = 0^+$ $\quad G(j\omega) = K\underline{/0^\circ}$

For $\omega \to +\infty$ $\quad G(j\omega) \to 0\underline{/-180^\circ}$

For negative frequencies the function is the conjugate of the above and the plot is a mirror image about the real axis. As there are no poles of $G(s)$ in the positive half-plane, for all values of K, τ_1 and τ_2 the -1 point is not enclosed and the system is always stable. It will be shown later that the degree of stability does vary with K.

$$(b) \quad G(s) = \frac{K}{s(1 + s\tau_1)(1 + s\tau_2)}$$

This function has a pole at the origin and the path on the s plane must be modified as shown in Fig. 4.7 (*a*). An enlarged plot of the region near $s = 0$ is shown in Fig. 4.7 (*b*).

Consider the plot of $G(j\omega)$ as ω varies from 0^+ to $+\infty$. This is shown in Fig. 4.8 (*a*) together with its mirror image from $\omega = 0^-$ to $-\infty$. Important points are

$\omega \to 0$ $\quad G(j\omega) \to \infty\underline{/-90^\circ}$

$\omega \to \infty$ $\quad G(j\omega) \to 0\ \underline{/-270^\circ}$

Referring to Fig. 4.8 (*a*), we have not yet completed the contour corresponding to the region $s = j0^-$ to $s = j0^+$ in Fig. 4.7 (*a*). This is completed by choosing a semi-circular path $s = \varepsilon e^{j\theta}$, where $\varepsilon \to 0$ (Fig. 4.7 (*b*)). Hence as

$$s \to 0 \qquad G(s) \to \frac{K}{s} = \frac{K}{\varepsilon e^{-j\theta}}$$

$$G(s) \to \frac{Ke^{j\theta}}{\varepsilon}$$

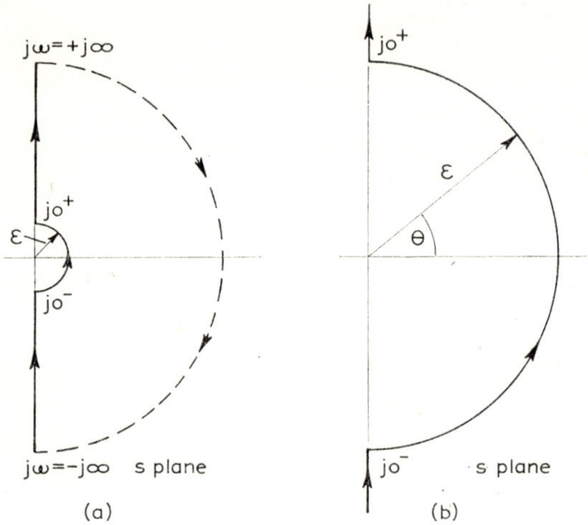

FIG 4.7

The magnitude of $G(s) \to \infty$ as $\varepsilon \to 0$ and its angle changes from $+\pi/2 \to -\pi/2$ as θ changes from $-\pi/2$ to $+\pi/2$. Thus the plot in Fig. 4.8 (*a*) is completed by joining the points $\omega = 0^-$ to $\omega = 0^+$ by a semicircle of infinite radius in a clockwise direction as shown in Fig. 4.8 (*b*). In this instance the system is stable since the -1 point is not enclosed and by inspection of $G(s)$ the number of poles in the right-hand half-plane, $P = 0$. If K is increased the -1 point will be enclosed by two clockwise rotations as ω goes from $-\infty$ to $+\infty$. Therefore $N = -2$, thus $Z = P - N = 2$ and the system is unstable, the characteristic equation having two zeros in the right-hand half-plane (see Example 5.1).

$$(c) \quad G(s) = \frac{K}{s^2(1 + s\tau_1)}$$

This has a double pole at the origin and a contour has to be chosen as in part (*b*). In this case the plot of $G(s)$ is completed from $\omega = 0^-$ to $\omega = 0^+$ by a rotation of 2π in a clockwise direction. The completed

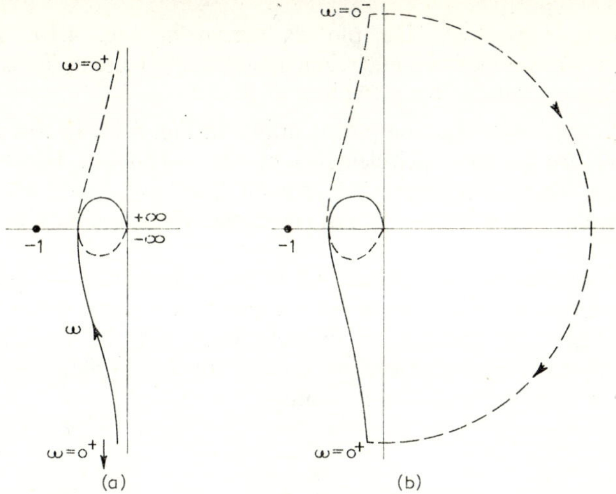

FIG 4.8

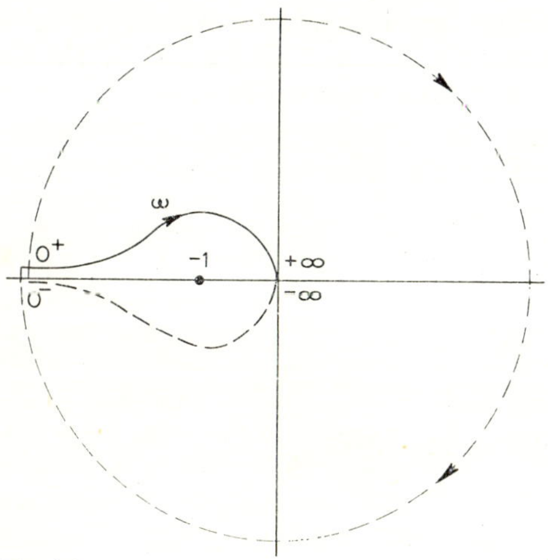

FIG 4.9

plot is shown in Fig. 4.9, giving $N = 2$ and as $P = 0$, $Z = 2$. The system is always unstable.

(*d*) $G(s) = \dfrac{K(1 + s\tau_1)(1 + s\tau_2)}{s^3(1 + s\tau_3)(1 + s\tau_4)}$

The shape of this plot can vary considerably depending on the relative magnitudes of τ_1, τ_2, τ_3 and τ_4. Two forms are given, but these again can vary in detail according to the actual values of the time constants.

1. $\tau_1 < \tau_2 < \tau_3 < \tau_4$. The plot is drawn in Fig. 4.10. This exhibits two clockwise encirclements of the -1 point. Thus the system is unstable for all values of $K > 1$.
2. $\tau_1 > \tau_2 > \tau_3 > \tau_4$. The plot is drawn in Fig. 4.11. In this case there are no net encirclements of the -1 point. However, increasing or diminishing the value of K will give two clockwise encirclements of the -1 point. Hence the system is conditionally stable.

   Note that if a radial line AB is drawn from the -1 point, inspection of the direction of the locus as it crosses this line indicates the number of encirclements; e.g. in the above example the line is crossed by two loci from left to right and two in the opposite direction indicating $N = 0$.

Example 4.2

Discuss the mathematical requirements for stability in a linear feedback system and give the Routh stability criterion. The closed-loop transfer function of a system is given by

$$\frac{\theta_o}{\theta_i} = \frac{50}{s(1 + s\tau)(1 + 0{\cdot}5s) + 50}$$

Determine (*a*) the value of τ such that the system just becomes unstable, and (*b*) the resulting frequency of oscillation. (*Manchester Univ.*)

Solution

The characteristic equation is

$$s(1 + s\tau)(1 + 0{\cdot}5s) + 50 = 0$$

after rearranging this becomes

$$\tau s^3 + (1 + 2\tau)s^2 + 2s + 100 = 0$$

The Routhian array is

$$\begin{array}{c|cc} s^3 & \tau & 2 \\ s^2 & (1 + 2\tau) & 100 \\ s^1 & \dfrac{2(1 + 2\tau) - 100\tau}{1 + 2\tau} & 0 \\ s^0 & 100 & \end{array}$$

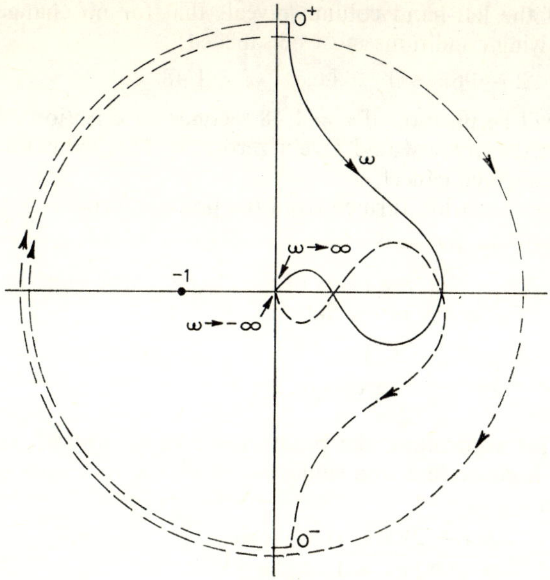

FIG 4.10

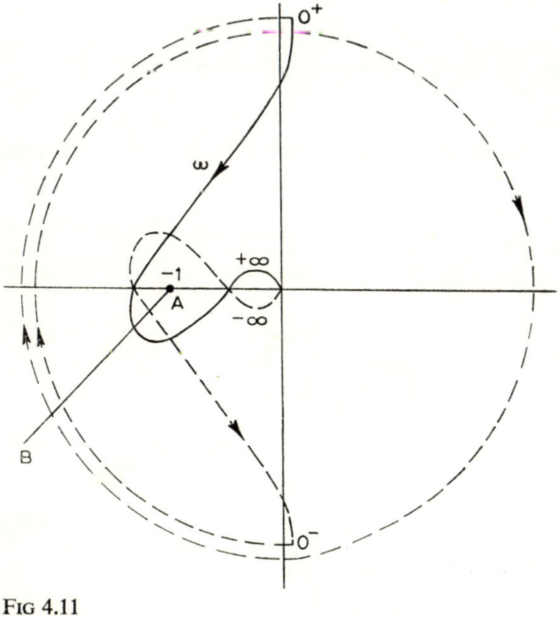

FIG 4.11

Inspection of the left-hand column reveals that for no changes of sign the following conditions must be satisfied:

$$\tau > 0 \qquad 2 - 96\tau > 0 \qquad \text{i.e.} \qquad \tau < 1/48$$

The system will be unstable if $\tau = 1/48$ second. Substitution of this value for τ in the array would give a zero row. This indicates that there are roots which cancel.

The auxiliary equation formed from the preceding row is

$$(1 + 1/24)s^2 + 100 = 0$$

giving $s = \pm j9{\cdot}8$, which are roots of the characteristic equation.

The angular frequency of oscillation is 9·8 rad/s.

Example 4.3

The system for controlling the performance of an aircraft in the lateral mode is equivalent to a single loop with the following open-loop function:

$$G(s)\,H(s) = \frac{K(s + 2)(s + 1)}{(s + 0{\cdot}1)(s - 1)}$$

Draw the complete direct Nyquist locus for the above system and determine the range of the gain constant K for which the system is stable.

Comment on the possible application of Bode's stability criterion to a Bode plot for the above system. (*Nottingham Univ.*)

Solution

The complete Nyquist locus is shown in Fig. 4.12. To obtain this locus use is made of the complex-number tables given in the appendix. Hence $G(s)\,H(s)$ is rewritten as

$$\frac{20K(1 + 0{\cdot}5s)(1 + s)}{(1 + 10s)(-1 + s)}$$

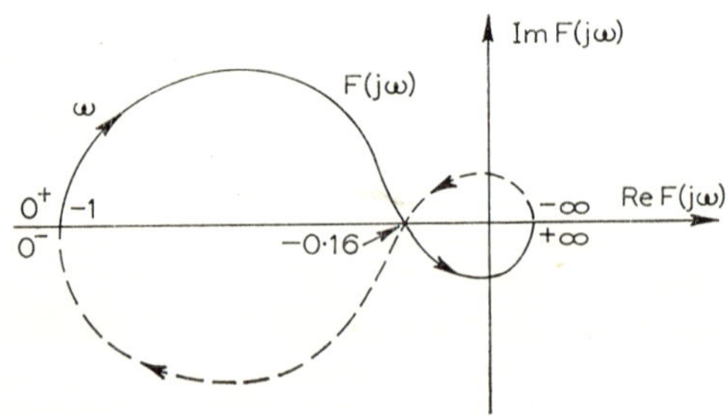

FIG 4.12

The values of the frequency-dependent part of the above expression, i.e.

$$F(j\omega) = \frac{(1 + 0{\cdot}5j\omega)(1 + j\omega)}{(1 + 10j\omega)(-1 + j\omega)}$$

are tabulated for positive values of ω and are plotted in Fig. 4.12.

ω (rad/s)	0	0·1	0·2	0·4	0·6
$F(j\omega)$	$1\underline{/180^\circ}$	$0{\cdot}71\underline{/148^\circ}$	$0{\cdot}45\underline{/144^\circ}$	$0{\cdot}25\underline{/158^\circ}$	$0{\cdot}17\underline{/183^\circ}$
ω (rad/s)	0·8	1·0	2·0	4·0	∞
$F(j\omega)$	$0{\cdot}13\underline{/195^\circ}$	$0{\cdot}11\underline{/202^\circ}$	$0{\cdot}07\underline{/265^\circ}$	$0{\cdot}05\underline{/307^\circ}$	$0{\cdot}05\underline{/0^\circ}$

The plot is completed for negative frequencies by reflecting the positive plot about the real axis. The locus of Fig. 4.12 must be scaled up by the factor $20K$ to obtain the true locus of $G(j\omega)\,H(j\omega)$.

Depending upon the value of K the -1 point may lie inside either the larger or the smaller loop of the locus. As K is increased the -1 point effectively moves to the right along the real axis.

If K is such that the critical point lies inside the larger loop, it is found by taking a radius vector from this point that the number of encirclements N is -1; i.e. one clockwise rotation is experienced. As there is a pole of $G(s)\,H(s)$ in the right-hand half-plane, $P = 1$, so that $-1 = 1 - Z$, or $Z = 2$. This indicates that there are two zeros of the function having positive real parts.

On the other hand, if the -1 point lies within the smaller loop $N = 1$ and $1 = 1 + Z$, so that $Z = 0$. There are no zeros having positive real parts and the system is stable. Thus the system will be on the point of stability if the -1 point coincides with the intersection of the two loops. This corresponds approximately to the point 0·16 on the uncorrected plot:

$$0{\cdot}16 = \frac{1}{20K} \qquad \text{giving} \qquad K = 0{\cdot}3$$

Therefore the system is stable if $K > 0{\cdot}3$. The angular frequency of oscillation would be approximately 0·58 rad/s at this value of K.

It is interesting to note that, if K is less than unity, i.e. the -1 point lies to the left of the complete locus, then $N = 0$ and we find that $Z = 1$, i.e. there is one zero with a positive real part. The point of transition from one zero to two zeros having positive real parts is when the magnitude of $KF(j\omega)$ at $\omega = 0$ is unity; i.e.

$$20K = 1 \qquad \text{giving} \qquad K = 0{\cdot}05$$

As a check on the above plot we perform the Routh stability test on the system. The characteristic equation is

$$1 + \frac{K(s + 2)(s + 1)}{(s + 0{\cdot}1)(s - 1)} = 0$$

giving

$$s^2(K + 1) + (3K - 0{\cdot}9)s + (2K - 0{\cdot}1) = 0$$

The array is

$$\begin{array}{c|cc} s^2 & K + 1 & 2K - 0{\cdot}1 \\ s^1 & 3K - 0{\cdot}9 & 0 \\ s^0 & 2K - 0{\cdot}1 & 0 \end{array}$$

From the left-hand column we obtain the following information:

1. There must be no changes of sign for stability, hence $K > 0{\cdot}3$ (we could have used the fact that all the coefficients of the original equation must be positive for stability).
2. If $K < 0{\cdot}05$ there would be one change of sign indicating one zero having a positive real part.
3. If $0{\cdot}05 < K < 0{\cdot}3$ there are two changes of sign and hence there are two zeros in the right-hand half of the s plane.

This example is solved using the root-locus technique in Example 5.4.

Example 4.4

State the Nyquist stability criterion and give a physical explanation justifying its validity. The open-loop transfer function of a control system is given by the following expression:

$$G(s) = \frac{K}{\tau s(1 + \tau s)(1 + \tau k s)}$$

Demonstrate that the maximum value of K to ensure stability is the same if obtained by the Nyquist stability criterion as by the Routh–Hurwitz stability criterion. Assume one feedback only with unity gain. (*Nottingham*)

Solution

Routh Criterion

The characteristic equation is

$$\frac{K}{\tau s(1 + \tau s)(1 + \tau k s)} + 1 = 0$$

$$k\tau^3 s^3 + (1 + k)\tau^2 s^2 + \tau s + K = 0$$

The Routh array is

$$\begin{array}{c|cc} s^3 & k\tau^3 & \tau \\ s^2 & (1 + k)\tau^2 & K \\ s & \dfrac{(1 + k)\tau^3 - Kk\tau^3}{(1 + k)\tau^2} & \end{array}$$

The limit of stability is given by

$$(1 + k)\tau^3 = Kk\tau^3$$

Therefore

$$K = \frac{1 + k}{k}$$

Nyquist Plot

The open-loop transfer function is

$$G(j\omega) = \frac{K}{j\omega\tau(1 + j\omega\tau)(1 + j\omega k\tau)}$$

and its form is plotted in Fig. 4.13.

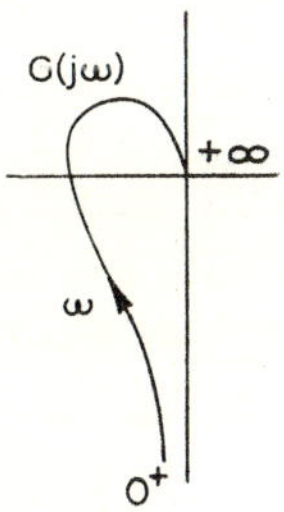

FIG 4.13

At the limit of stability, therefore,

$$\frac{K}{j\omega\tau(1 + j\omega\tau)(1 + j\omega k\tau)} = -1$$

$$-K = j\omega\tau(1 + j\omega\tau[1 + k] - \omega^2\tau^2 k)$$

Equating real and imaginary parts,

$$K = \omega^2\tau(1 + k) \tag{1}$$

and

$$0 = \omega\tau - \omega^3\tau^3 k$$

whence $\omega = 0$, or $\omega^2 = 1/\tau k$.

Substituting for ω in eqn. (1),

$$K = (1 + k)/k$$

Example 4.5

Show that the network in Fig. 4.14 has a transfer function of the form

$$\frac{V_o}{V_i} = \frac{a(1 + s\tau)}{1 + as\tau}$$

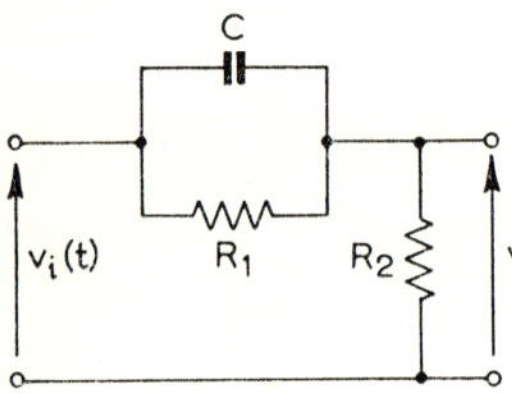

FIG 4.14

What values of R_1 and R_2 will give $\tau = 0{\cdot}5$ second and $a = 0{\cdot}1$ if $C = 1\,\mu F$?

In the system shown in Fig. 4.15, $K = 1{\cdot}6$, $\tau = 0{\cdot}1$ second and $G(s) = 1$. What is the maximum value of k for stability?

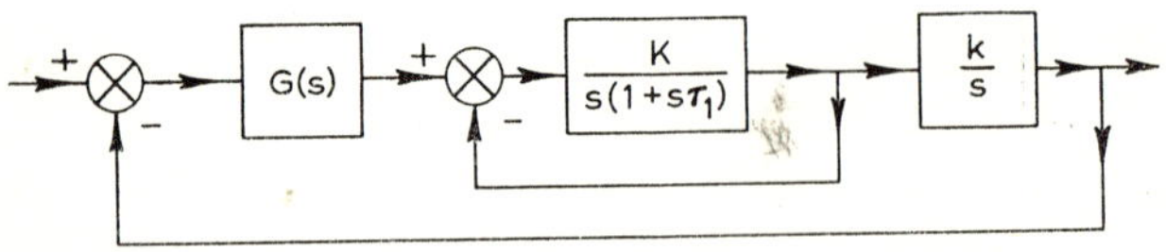

FIG 4.15

If $G(s)$ is changed to the network of Fig. 4.14, what then is the maximum value of k, assuming the network constants stipulated? (*Loughborough*)

Solution

To show that

$$\frac{V_o}{V_i} = \frac{a(1 + s\tau)}{(1 + as\tau)}$$

see Example 1.1 (*c*), where $\tau = CR_1$, and $a = R_2/(R_1 + R_2)$. When $a = 0{\cdot}1$ and $\tau = 0{\cdot}5$ s,

$$C = 1\,\mu\text{F} \qquad R_1 = 0{\cdot}5\,\text{M}\Omega \qquad R_2 = 0{\cdot}055\,\text{M}\Omega$$

The characteristic equation of the system shown in Fig. 4.15 when $G(s) = 1$ is

$$\frac{k}{s}\left[\frac{K/s(1 + s\tau_1)}{1 + K/s(1 + s\tau_1)}\right] + 1 = 0$$

$$s^3\tau_1 + s^2 + sK + kK = 0$$

Substituting $\tau_1 = 0{\cdot}1$ and $K = 1{\cdot}6$,

$$0{\cdot}1s^3 + s^2 + 1{\cdot}6s + 1{\cdot}6k = 0$$

The Routhian array is

$$\begin{array}{c|cc} s^3 & 0{\cdot}1 & 1{\cdot}6 \\ s^2 & 1 & 1{\cdot}6k \\ s^1 & \dfrac{1{\cdot}6 - 0{\cdot}16k}{1} & \end{array}$$

The limit of stability is reached when $k = 10$. When

$$G(s) = \frac{0{\cdot}1(1 + 0{\cdot}5s)}{(1 + 0{\cdot}05s)}$$

the characteristic equation becomes

$$\frac{0{\cdot}1(1 + 0{\cdot}5s)kK}{(1 + 0{\cdot}05s)s(s[1 + s\tau_1] + K)} + 1 = 0$$

Inserting the numerical values,

$$0{\cdot}005s^4 + 0{\cdot}15s^3 + 1{\cdot}08s^2 + (1{\cdot}6 + 0{\cdot}08k)s + 0{\cdot}16k = 0$$

The new array is

$$\begin{array}{c|ccc} s^4 & 0{\cdot}005 & 1{\cdot}08 & 0{\cdot}16k \\ s^3 & 0{\cdot}15 & 1{\cdot}6 + 0{\cdot}08k & 0 \\ s^2 & \dfrac{0{\cdot}162 - 0{\cdot}005(1{\cdot}6 + 0{\cdot}08k)}{0{\cdot}15} & 0{\cdot}16k & 0 \\ s^1 & \dfrac{A(1{\cdot}6 + 0{\cdot}08k) - 0{\cdot}024k}{A} & 0 & 0 \end{array}$$

where

$$A = \frac{0{\cdot}162 - 0{\cdot}005(1{\cdot}6 + 0{\cdot}08k)}{0{\cdot}15}$$

For a change in sign to occur in the coefficient of s^2,

$$0{\cdot}162 - (8 \times 10^{-3}) = 4 \times 10^{-4}k \qquad \text{so that} \qquad k = 385$$

For a change in sign to occur in the coefficient of s,

$$\left[\frac{0{\cdot}162 - 0{\cdot}005(1{\cdot}6 + 0{\cdot}08k)}{0{\cdot}15}\right](1{\cdot}6 + 0{\cdot}08k) - 0{\cdot}024k = 0$$

$$(-0{\cdot}32 \times 10^{-4}k^2) + (0{\cdot}806 \times 10^{-2}k) + 0{\cdot}246 = 0$$

The solution to this equation is $k = 27$, and this is the maximum value of k for stability.

PROBLEMS

4.1 A closed-loop control system has an open-loop transfer function given by

$$KG(s) = \frac{\omega_0}{s(1 + \tau s)^2}$$

If $\tau = 0{\cdot}1$ second, plot the Nyquist diagram and so find the maximum value of ω_0 for stability. (*Brunel Univ.*)

Ans. 20.

4.2 Explain and justify the Nyquist criterion of stability when applied to a feedback system having only one feedback loop.

Sketch the transfer locus for the open-loop transfer function

$$G(s) = \frac{K}{s(\tau_1 s + 1)(\tau_2 s + 1)}$$

and explain the effect of the value of K on the stability of the system when the loop is closed. What effect will the ratio of τ_1 to τ_2 have on system stability and how will it affect the value of K? (*Cardiff*)

Ans. For stability, $0 < K < (1/\tau_1 + 1/\tau_2)$.

4.3 Explain what is meant by the limit of stability of a linear system and describe how the stability of such a system can be checked (*a*) on a Nyquist diagram, and (*b*) using the Routh–Hurwitz criterion. Briefly justify your arguments.

Explain why sometimes integrations are added in the forward path of the system and comment on their effect on the system stability.

A particular system has the open-loop transfer function

$$\frac{K}{s(1 + \tau s)^2}$$

Indicate the Nyquist diagram and compute the maximum value of K for stability. (*Brunel Univ.*)

Ans. $K = 2/\tau$.

4.4 Describe some criterion for ascertaining whether the roots of a polynomial all lie in the left-half complex plane. Given the polynomial

$$s^5 + 2s^4 + 10s^3 + (8 + A)s^2 + 9s + 4A$$

determine the range of values of A such that its zeros all lie in the left-half complex plane. (*I.E.R.E.*)

Ans. $0 < A < 0{\cdot}2$.

4.5 Describe an experiment to determine the open-loop frequency response of a type 1 position control system. A type 0 system has an open-loop transfer function

$$A(1 + 0{\cdot}01s)/(1 + s)^2(1 + 0{\cdot}1s)$$

Determine the value of A at which the system becomes unstable. (*Southampton Univ.*)

Ans. $A = 27{\cdot}3$.

4.6 Outline the proof of the generalized Nyquist stability criterion. Determine the stability of each of the four systems whose closed-loop transfer possess the denominators given below:

(i) $s^3 - s^2 + s + 1$
(ii) $4s^4 - 20s^3 + 33s^2 + 32s + 15$
(iii) $4s^4 + 8s^3 + s^2 + 7s + 5$
(iv) $s^4 + 7s^3 + s + 3$

Give full reasons for your conclusion in each case. If (iii) represented the denominator of the product $H(s)\,G(s)$ rather than the denominator of the closed-loop transfer function, state what further information would be required to ascertain the stability of the system. Briefly outline the procedure which would be involved in the stability determination. (*Southampton Univ.*)

Ans. All unstable.

4.7 State and explain the Nyquist stability criterion. Explain how a function with poles and zeros is transformed from the s plane to the y plane. Illustrate by plotting the poles and zeros of the transfer function

$$y(s) = \frac{(s-1)}{(s+1)}$$

in the s plane and transforming into the y plane.

If a feedback system has a forward transference

$$\frac{K(s-1)}{(s+1)}$$

and unity feedback, for what range of values of K is it stable? (*Swansea*)

Ans. $K < 1$.

4.8 Define the term "characteristic equation" as applied to closed-loop systems and show how it can be used to determine whether or not the system is stable. A system with unity feedback ratio has the following open-loop transfer function:

$$\frac{\theta_0(s)}{E(s)} = \frac{7\,500(s+K)}{s^2(s+34{\cdot}5)}$$

Determine a range of values of K such that the system is always stable. (*Nottingham Univ.*)

Ans. $0 < K < 34{\cdot}5$.

4.9 (*a*) State how it is possible to ascertain from the sketch of the Nyquist plot of a loop transference $G(s)H(s)$ just how many roots of $1 + GH = 0$ lie in the right-half s plane.

(*b*) If it is desirable that all roots of the characteristic equation $1 + GH(s) = 0$ have real parts less than $-\alpha$, what kind of Nyquist curve should be constructed. Demonstrate your answer with the function

$$GH = \frac{K}{(s+1)(s+2)(s+3)}$$

where $\alpha = 1$. Find the largest value of K satisfying the condition. (*Swansea*)

Ans. 6. *Hint*. In the s plane the line $s = -1 + j\omega$ is used instead of the imaginary axis ($s = j\omega$). To use Routh's criterion a new variable ($s' = s + 1$) is defined giving a new characteristic equation.

4.10 (*a*) Derive an expression for the general linearized transfer function of a field-controlled motor relating the armature position θ to the field voltage V_f in terms of the Laplace operator s and the following parameters:

J = moment of inertia of the motor armature
F = viscous frictional torque per unit angular velocity of motor armature
L_f = inductance of field winding
R_f = resistance of field winding
K = motor torque per unit field current

Assume that the armature current is effectively constant.

(*b*) A small instrument servomechanism for position control of a field-controlled d.c. motor is supplied by an electronic amplifier. The armature current of the motor is maintained constant and the electronic amplifier receives an error voltage which is derived from the error detector. The error detector is a potentiometer [voltage divider] bridge with a sensitivity of 1 V/rad of error. Other parameters of the system are:

Electronic amplifier. Gain into open-circuit, 100 V/V; output resistance, 100 Ω; input resistance, effectively infinite.

Motor. $R = 100\,\Omega$; $F = 2 \times 10^{-2}$ N-m/rad-s^{-1};

$J = 2 \times 10^{-3}$ kg-m^2; $K = 1{\cdot}0$ N-m; $L_f = 1$ H.

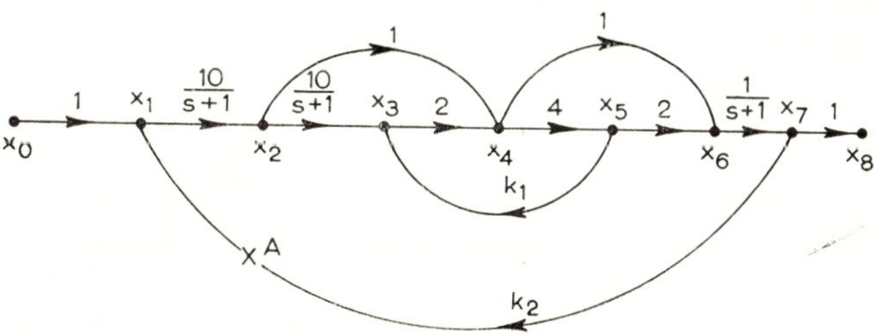

FIG 4.16

(i) Draw a simplified circuit diagram for the system.
(ii) Draw a block diagram and develop an expression for the overall general open-loop transfer function.
(iii) Determine analytically whether or not the system will be stable on closed-loop. (*I.E.R.E.*)

Ans. System is stable. $\dfrac{\theta_0}{\varepsilon} = \dfrac{25}{s(0{\cdot}005s + 1)(0{\cdot}1s + 1)}$

4.11 A closed-loop system is represented by the signal-flow diagram shown in Fig. 4.16. For what values of k_1 is the system stable with the main loop opened at A? If $k_1 = -1$ for what values of k_2 is the system stable on closed-loop? If $k_1 = -1$, $k_2 = -0{\cdot}1$, discuss the extent to which the three time-constants of the system could be changed to new but equal values without creating instability. (*Birmingham Univ.*)

Ans. $k_1 < 1/8$, $-2/45 < k_2 < 1/210$.

Chapter 5

The Root-locus Plot

Up to this point we have related the position on the s plane of the roots (i.e. the zeros) of the characteristic equation to system stability. These roots are the poles of the closed-loop transfer function and the *root-locus method* is a graphical approach to determining the position of these poles. The characteristic equation

$$1 + G(s)H(s) = 0$$

is satisfied when

$$|G(s)H(s)| = 1 \tag{5.1}$$

and

$$\underline{/G(s)H(s)} = 180^\circ \tag{5.2}$$

The process of finding points on the s plane which satisfy these conditions is the basis of the root-locus method.

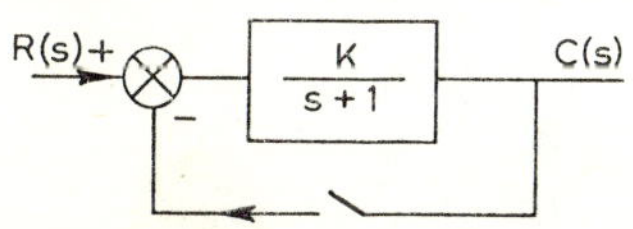

Fig 5.1

Consider the simple system depicted in Fig. 5.1. The open-loop transfer function

$$G(s)H(s) = \frac{K}{s+1}$$

has a pole at $s = -1$ (see Fig. 5.2). The closed-loop transfer function is

$$\frac{C}{R} = \frac{G(s)}{1 + G(s)H(s)} = \frac{K}{(s + K + 1)}$$

This has a pole at $s = -(K + 1)$, hence as K is increased from zero to infinity the closed-loop pole moves to the left along the real axis away from its open-loop position.

Note. A term of the form $A/(s + \sigma)$ on the s plane is transformed in the time domain to $A\,e^{-\sigma t}$, and the larger the value of σ the faster is the system response. Increasing the value of K, the open-loop gain, gives rise to a smaller time-constant, $1/(K + 1)$; see Example 2.2.

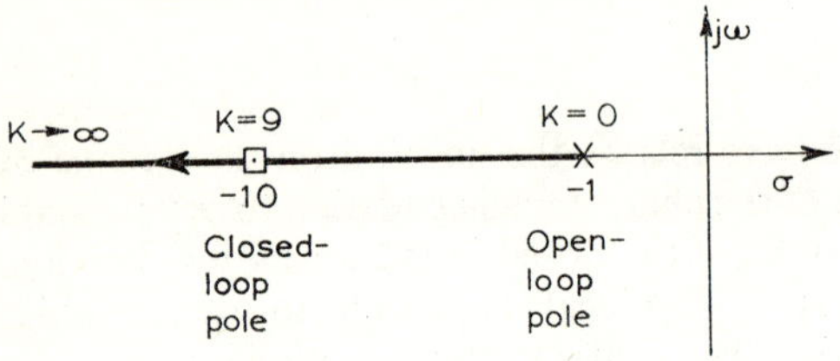

FIG 5.2 Path of pole of $\dfrac{C}{R} = \dfrac{K}{s + K + 1}$ as K increases

It is apparent that any point on the path shown in Fig. 5.2 satisfies the angle condition given in eqn. (5.2), i.e. any point on this path subtends an angle of 180° when viewed from the open-loop pole at $s = -1$. The actual position of the closed-loop pole on the locus is determined by the value of K (e.g. if $K = 9$ the pole would be at $s = -10$).

As a further example consider the voltage control system discussed in the introduction to Chapter 3. The block-diagram of the system is repeated in Fig. 5.3.

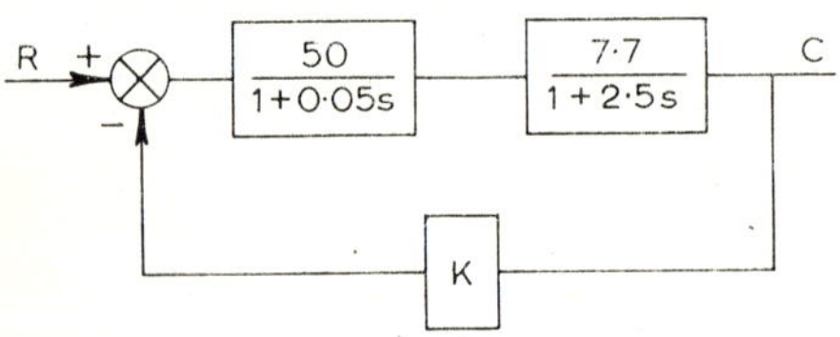

FIG 5.3

The open-loop transfer function is

$$\frac{(50)(7{\cdot}7)K}{(1 + 0{\cdot}05s)(1 + 2{\cdot}5s)}$$

which is rewritten as

$$\frac{K'}{(s + 20)(s + 0{\cdot}4)}$$

where $K' = 3080K$. This has poles at $s = -20$ and $s = -0{\cdot}4$.

In the closed-loop condition the poles are at

$$s = -6{\cdot}2 \text{ and } -14{\cdot}2 \text{ for } K = 0{\cdot}026$$
$$s = -10{\cdot}2 \text{ and } -10{\cdot}2 \text{ (a double pole) for } K = 0{\cdot}031$$
$$s = -10{\cdot}2 + j17{\cdot}5 \text{ and } -10{\cdot}2 - j17{\cdot}5 \text{ for } K = 0{\cdot}13$$

The corresponding values of K' are 80, 96 and 400.

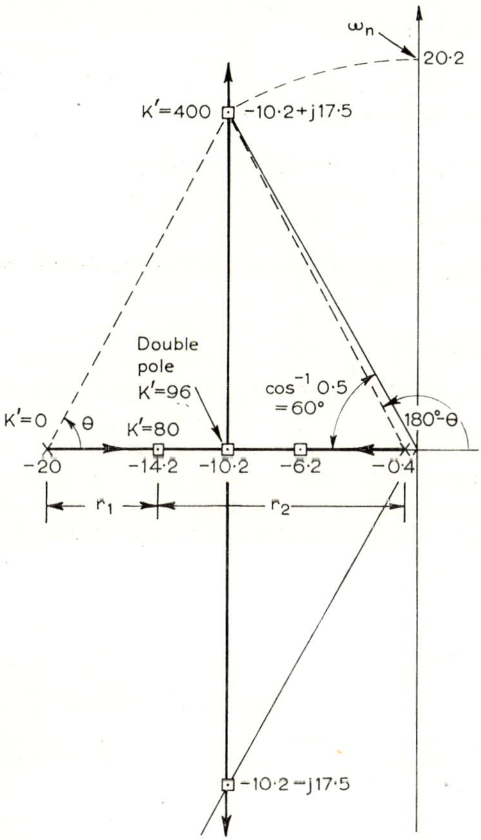

FIG 5.4

The pole positions are plotted in Fig. 5.4, together with the complete locus as K is varied. This locus satisfies the angle condition of eqn. (5.2).

For the position on the real axis the pole at $s = -0{\cdot}4$ contributes 180° whereas the pole at $s = -20$ contributes 0°. Simple geometry shows that a vertical line through the mid-point of the poles also satisfies this condition.

To determine the actual position of the closed-loop poles on the locus, use is made of eqn. (5.1). For $K' = 80$,

$$\left|\frac{80}{(s + 20)(s + 0{\cdot}4)}\right| = 1$$

i.e.

$$|(s + 20)(s + 0{\cdot}4)| = 80 \tag{5.3}$$

As we already know the position of the closed-loop poles for $K' = 80$, eqn. (5.3) can be checked. For $s = -6{\cdot}2$,

$$|(s + 20)(s + 0{\cdot}4)| = |(20 - 6{\cdot}2)(0{\cdot}4 - 6{\cdot}2)| = 80$$

Similarly for $s = -14{\cdot}2$.

For $K' = 96$, $s = -10{\cdot}2$ giving $|(20 - 10{\cdot}2)(-10{\cdot}2 + 0{\cdot}4)| = 96$.
For $K' = 400$, $s = -10{\cdot}2 \pm j17{\cdot}5$.

Taking $s = -10{\cdot}2 + j17{\cdot}5$, we have

$$|(s + 20)(s + 0{\cdot}4)| = |(9{\cdot}8 + j17{\cdot}5)(-9{\cdot}8 + j17{\cdot}5)| = 400$$

An alternative method would be to measure the distances r_1 and r_2 to scale on Fig. 5.4. Multiplication of these quantities gives K'.

The form of the time response with a pair of complex poles can be obtained by writing the characteristic equation in the standard form

$$\frac{C}{R}(s) = \frac{\omega_n^2}{s^2 + 2\zeta\omega_n s + \omega_n^2}$$

This can be written

$$\frac{C}{R}(s) = \frac{\omega_n^2}{[s + \zeta\omega_n + j\omega_n\sqrt{(1 - \zeta^2)}][s + \zeta\omega_n - j\omega_n\sqrt{(1 - \zeta^2)}]}$$

which has poles at

$$s = -\zeta\omega_n \pm j\omega_n\sqrt{(1 - \zeta^2)}$$

The variation of $c(t)$ following a step input has already been derived as eqn. (3.6):

$$c(t) = 1 - \frac{e^{-\zeta\omega_n t}}{\sqrt{(1 - \zeta^2)}} \sin[\omega_n\sqrt{(1 - \zeta^2)}t + \cos^{-1}\zeta]$$

The position of the complex poles of $\frac{C}{R}(s)$ can be interpreted according to the oscillatory term in this equation.

1. Lines of constant $\zeta\omega_n$ are vertical, and represent lines of constant decay.
2. Lines of constant $\omega_n\sqrt{(1 - \zeta^2)}$ are horizontal, and represent lines of constant damped natural frequency ω_d.
3. Circles of constant undamped natural frequency ω_n (centre the origin) have radii given by

$$r^2 = (\zeta\omega_n)^2 + \omega_n^2(1 - \zeta^2) = \omega_n^2$$

4. Lines of constant angle η to the real axis have angle η given by

$$\cos^{-1} \eta = \zeta\omega_n/\omega_n = \zeta$$

These are lines of constant damping ratio ζ.

The examples used have been simple enough to calculate the position of the closed-loop poles, and eqns. (5.1) and (5.2) have been used only as a check. In general, eqns. (5.1) and (5.2) are used to determine the root locus and the position of the poles. To facilitate the construction of the locus the following rules are available.

Rule 1. The number of loci is equal to the order of the characteristic equation.

Rule 2. Each locus starts at an open-loop pole when $K = 0$ and finishes either at an open-loop zero (or infinity) when $K = \infty$.

Rule 3. Loci either run along the real axis or occur as complex conjugate pairs.

Rule 4. A point on the real axis is part of the loci if the sum of the poles and zeros to the right of the point concerned is odd.

Rule 5. As one moves far enough away from the open-loop poles and zeros the root loci become asymptotic to lines making angles α_n to the real axis:

$$\alpha_n = \pm \frac{n\pi}{P - Z}$$

P being the number of open-loop poles, Z the number of open-loop zeros, and n an odd integer.

Rule 6. The asymptotes intersect the real axis at a point s given by

$$s = \frac{\sum \text{poles} - \sum \text{zeros}}{P - Z}$$

where P is the number of open-loop poles, Z the number of open-loop zeros, and s the centroid of the open-loop pole-zero configuration.

Rule 7. The break-away point between two poles or break-in point between two zeros can be obtained by the following method. Consider Fig. 5.4: here we see that the closed-loop poles move along the real axis as K' is increased. The maximum value of K' obtained before the poles leave the real axis is 96. Such a point is called a *break-away point*. The condition satisfied by the position of a closed-loop pole is $K'G(s) = -1$. For convenience we alter the form slightly to

$$\frac{K'}{W(s)} = -1 \qquad \text{i.e.} \qquad W(s) = -K'$$

We may now investigate the variation of K' with complex frequency. Here we are interested in the maximum value of K' for real values of s (i.e. $s = \sigma$). Hence by differentiating $W(s)$ we can obtain the maximum (break-away point) or minimum (break-in point) of the function along the real axis.

Rule 8. The crossing points of the imaginary axis can be obtained by Routh's criterion.

Rule 9. The angle of departure from a complex pole or zero is given by $180° - \phi_p + \phi_z$, where ϕ_p is the sum of all the angles subtended by the other poles (including the conjugate pole), and ϕ_z is the sum of the angles of any zeros.

Example 5.1

Sketch the root-locus diagram for a control system with unity feedback having the forward-path transfer function

$$\frac{K}{s(s + \sigma_1)(s + \sigma_2)}$$

where $\sigma_1 = 2s^{-1}$, and $\sigma_2 = 5s^{-1}$.

Find the least value of K to give an oscillatory response and the greatest value of K which can be used before continuous oscillation occurs. Find the frequency of the continuous oscillation when K is just large enough to give this condition.

Obtain the value of K to give a closed-loop response dominated by a pair of complex poles with damping ratio of 0·5. Hence obtain, in factored form suitable for partial-fraction expansion, the closed-loop transfer function corresponding to this value of K. (*Brunel Univ.*)

Solution

The function $K/s(s+2)(s+5)$ has open-loop poles at 0, -2 and -5 on the complex plane (see Fig. 5.5). As the characteristic equation is of the third order, i.e.

$$\frac{K}{s(s+2)(s+5)} = -1$$

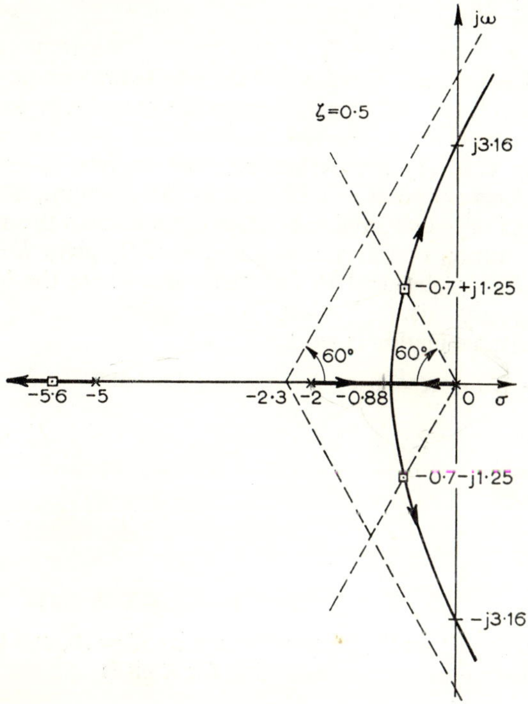

FIG 5.5

there are three branches of the root locus (Rule 1), each branch describing the variation of a particular closed-loop pole with the variable K. When $K = 0$ the open- and closed-loop poles coincide (Rule 2). Using Rule 4, there is a branch (or segment) of the locus along the real axis between the origin and -2, and a further branch between -5 and $-\infty$.

The angles of the asymptotes of the locus as $s \to \infty$ are (by Rule 5) $\alpha_1 = \pm \pi/3 = \pm 60°$, using a value of $n = 1$. Putting $n = 3$ would give $\alpha_3 = \pm \pi$, i.e. real axis as above.

The point of intersection of the asymptotes and the real axis is given by Rule 6. This occurs at

$$\frac{-(5+2+0)}{3} = -2{\cdot}33$$

The break-away point from the real axis can be found by noting that

$$|K| = |s(s+2)(s+5)| \tag{1}$$

As $K = 0$ at $s = 0$ and $s = -2$, it must reach a maximum at a point between these values. This point is the break-away point, and here the locus leaves the real axis and enters the complex region as two branches.

Differentiating eqn. (1) and setting it equal to zero gives the maximum (break-away) point or minimum (break-in) point, if any. Hence $s = -3{\cdot}85$ or $-0{\cdot}88$, and the latter value satisfies the angle condition. Substitution of this value of s in eqn. (1) gives $K = 4$. Routh's rule can now be used to find the point where the locus crosses the imaginary axis.

The characteristic equation is written

$$s^3 + 7s^2 + 10s + K = 0$$

and the array is,

$$\begin{array}{c|cc} s^3 & 1 & 10 \\ s^2 & 7 & K \\ s^1 & \dfrac{(70-K)}{7} & 0 \\ s^0 & K & \end{array}$$

For the system to be on the verge of instability $K = 70$, and from the preceding line in the array substitution for K gives

$$7s^2 + 70 = 0 \qquad \text{i.e.} \qquad s = \pm j3{\cdot}16\,\text{rad/s}$$

The complete plot is shown in Fig. 5.5.

The condition for a pair of dominant poles to have a damping ratio of 0·5 is satisfied at the intersection of the locus and a line from the origin at an angle of 60° to the negative real axis. By measurement, the value of K for this condition is $1{\cdot}45 \times 1{\cdot}8 \times 4{\cdot}45 = 11{\cdot}5$. The complex closed-loop poles for this new value of K are $-0{\cdot}7 \pm j1{\cdot}25$.

The pole on the negative real axis is found by factorizing the cubic equation

$$s^3 + 7s^2 + 10s + 11{\cdot}5 = 0$$

knowing that a factor is $(s^2 + 1{\cdot}4s + 2{\cdot}05)$ due to the complex poles.

The closed-loop transfer function is

$$\frac{11{\cdot}5}{(s + 5{\cdot}6)(s + 0{\cdot}7 + j1{\cdot}25)(s + 0{\cdot}7 - j1{\cdot}25)}$$

Example 5.2

Explain briefly what is meant by a root locus with reference to a closed-loop control system. The open-loop transfer function of a control system is given:

$$G(s) = \frac{K}{s(s^2 + 4s + 13)}$$

Draw the root loci and determine the value of K if the damping ratio of the oscillatory mode is to be about 0·5. Also find the natural frequency.

What form would the other terms take in the response for an assumed step-function input? (*Cardiff*)

Solution

The open-loop transfer function has poles at $s = 0$, $s = -2 \pm j3$. Application of the rules listed in the introduction give the following properties of the plot.

There are three branches, starting for $K = 0$ at $s = 0$, $s = -2 + j3$ and $s = -2 - j3$. One branch is the real axis to left of the origin. The asymptotes are at the following angles to the real axis:

$$\alpha = \pm n\pi/3 = \pm 60° \text{ and } 180° \text{ for } n = 1 \text{ and } n = 3$$

Intersection of the asymptotes and the real axis occurs at

$$s = \frac{0 + (-2 + j3) + (-2 - j3)}{3} = -4/3$$

Routh's criterion gives the crossing points on the imaginary axis $s = \pm j3{\cdot}6$ at $K = 52$.

The angle of departure ψ, of the locus from the complex pole $s = -2 + j3$ is

$$\psi = 180° - 90° - 125° = -35°$$

Similarly for the pole at $s = -2 - j3$, $\psi = 35°$.

The root-locus plot is shown in Fig. 5.6. For a damping ratio of 0·5 a line is drawn from the origin at an angle of $\cos^{-1}(0{\cdot}5) = 60°$ to the negative real axis. The root locus crosses this line at $s = -1{\cdot}6 \pm j2{\cdot}8$, and by measurement $K = 8$ at this point. Direct measurement of the distance of this point from the origin gives $\omega_n = 3{\cdot}2$ rad/s.

When $K = 8$ the open-loop pole at the origin has moved along the real axis to $s = -0{\cdot}75$ (this is obtained by measuring K along

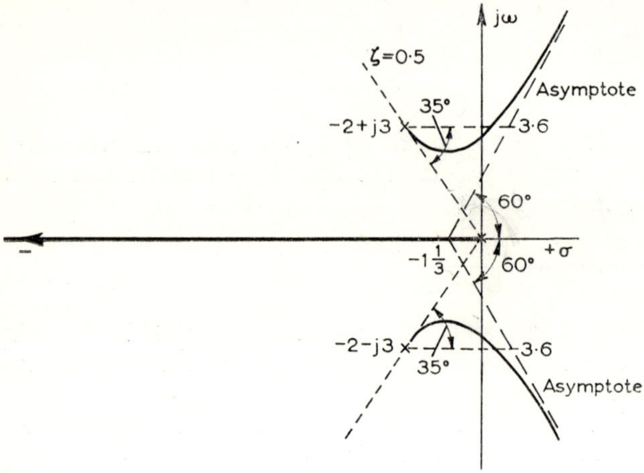

FIG 5.6

the real axis on a trial-and-error basis). The closed-loop transfer function for this value of K takes the form

$$\frac{C}{R}(s) = \frac{K'}{(s + 0{\cdot}75)(s + 1{\cdot}6 + j2{\cdot}8)(s + 1{\cdot}6 - j2{\cdot}8)}$$

and the response to a unit step input $R(s) = 1/s$ is

$$C(s) = \frac{K'}{s(s + 0{\cdot}75)(s + 1{\cdot}6 + j2{\cdot}8)(s + 1{\cdot}6 - j2{\cdot}8)}$$

Besides the oscillatory term there are two other terms in the time response. These are

$$K_1 \mathrm{e}^{-t} = K_1 \qquad \text{and} \qquad K_2 \mathrm{e}^{-0{\cdot}75t}$$

where K_1 and K_2 are constants. These constants can be evaluated by measurements on the root-locus plot (see later examples).

Example 5.3

The open-loop transfer function of a control system is of the form

$$G(s) = \frac{K(s + a)}{s^2(s + b)}$$

Show the way in which the root loci change as a varies over the range from zero to b.

If $a = 2$ and $b = 10$ sketch the root locus and determine approximately the value of K which gives the greatest damping ratio for the oscillatory mode. Find also the value of the damping ratio and the natural frequency. (*Cardiff*)

Solution

If $a = 0$,

$$G(s) = \frac{K}{s(s+b)}$$

which gives the root locus of Fig. 5.7 (a). If $a = b$,

$$G(s) = \frac{K}{s^2}$$

giving the locus of Fig. 5.7 (c). In between these values the locus is represented by Fig. 5.7 (b).

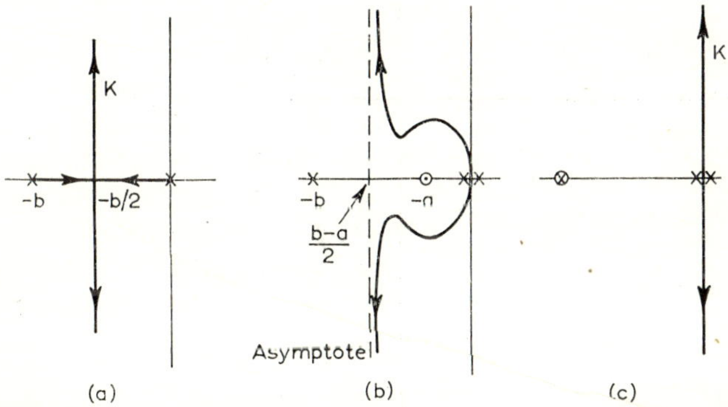

FIG 5.7

In the general case where $0 < a < b$ the asymptotes are

$$\pm\frac{180^\circ}{(3-1)} = \pm 90^\circ$$

(There are three poles and one zero.)

The centroid of the pole zero pattern, i.e. the intersection of the asymptotes and the real axis, is

$$s = -\frac{b + 2(0) - a}{2} = \frac{-(b-a)}{2}$$

When $a = b$ the zero cancels the pole on the real axis and the locus then lies along the imaginary axis. Substituting the values for a and b,

$$G(s) = \frac{(s + 2)}{s^2(s + 10)}$$

The centroid becomes $s = -4$ and the complete plot is shown in Fig. 5.8.

The system can never become unstable (this is easily confirmed by applying Routh's rule). A tangent from the origin to the locus (see

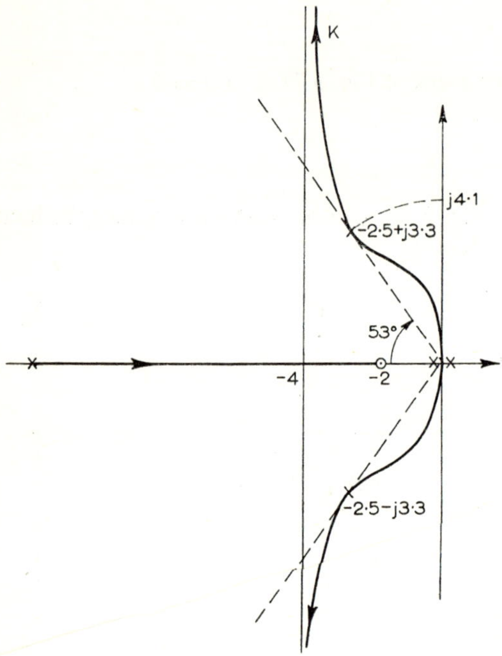

FIG 5.8

Fig. 5.8) gives the maximum value of ζ. By measurement for this value of damping ratio,

$$K = \frac{(4{\cdot}2)(8{\cdot}1)}{3{\cdot}3} = 43$$

The closed-loop poles are $-2{\cdot}5 \pm j3{\cdot}3$ for this condition, and the damped frequency of oscillation is thus 3·3 rad/s. By direct measurement the undamped natural frequency is 4·1 rad/s.

Example 5.4

A system for controlling the performance of an aircraft in the lateral mode is equivalent to a single loop with the following open-loop function:

$$G(s)\,H(s) = \frac{K(s + 2)(s + 1)}{(s + 0{\cdot}1)(s - 1)}$$

Draw the complete direct Nyquist locus for the above system and determine the range of gain constant for which the system is stable. Comment on the possible application of Bode's stability criterion to a Bode plot for the above system. (*Nottingham Univ.*)

Solution

This problem has already been solved in Chapter 4, but it is instructive to solve it again using root-locus techniques.

The function $G(s)\,H(s)$ is an example of a non-minimum phase function because it has a pole in the right-hand half-plane. The root-

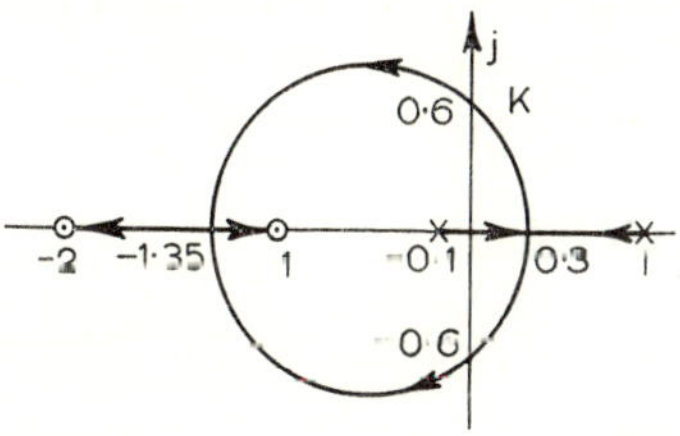

FIG 5.9

locus plot for variation in K is shown in Fig. 5.9. To find the break-away and break-in points write

$$W(s) = \frac{(s + 0{\cdot}1)(s - 1)}{(s + 2)(s + 1)}$$

Differentiation of this function would yield the maximum and minimum values of $W(s)$. However, it is easier to plot $|W(s)|$ against the real part of s as shown in Fig. 5.10. The break-away point from the real axis is at $s = 0{\cdot}3$ ($K = 0{\cdot}1$), the break-in point being at $s = -1{\cdot}35$ ($K = 12{\cdot}9$). From the root-locus plot the following information can be obtained concerning the stability of the system.

The system is unstable in the open-loop condition ($K = 0$) with one pole in the right-hand half-plane. As K is increased the open-loop pole at $s = -0{\cdot}1$ also moves into the right-hand half-plane.

By measurement this occurs when $K = 0{\cdot}05$. The system is now unstable with two poles in this half-plane.

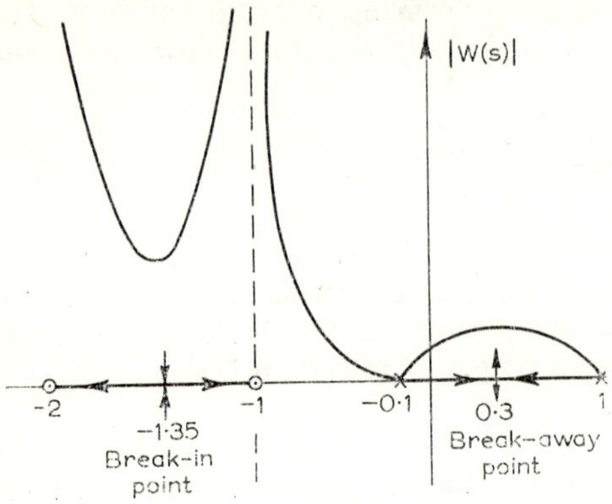

FIG 5.10

As K is increased further these poles move as shown and enter the left-hand half-plane when $K = 0{\cdot}3$ at $s = j0{\cdot}6$. The system is then stable for further increase in K.

Example 5.5

A proportional control system has a forward transfer function given by the expression

$$G(s) = \frac{\theta_0}{\theta}(s) = \frac{100}{s(0{\cdot}2s + 1)}$$

Determine the steady-state error of the system when subjected to a unit ramp function and by plotting the position of the poles of the closed-loop transfer function in the s plane estimate the transient behaviour of the system.

If a proportional-plus-derivative network having a transfer function of the form $(1 + 0{\cdot}1s)$ is inserted in the system, estimate the effect on (*a*) the steady-state error, (*b*) the transient performance of the system, when operated as in the previous case. (*Leeds Univ.*)

Solution

The forward transfer function is put in the form

$$G(s) = \frac{500}{s(s + 5)}$$

to emphasize the position of the poles. Assuming a unity feedback path, the root locus is as shown in Fig. 5.11 (*a*). The closed-loop poles are obtained by using the equation $|s(s+5)| = 500$; by symmetry, the distance to the locus from an open-loop pole to the closed-loop pole will be $\sqrt{500}$. The latter poles are at $s = -2{\cdot}5 \pm j22{\cdot}2$.

The expression for $\theta_0(s)$ when the system is subjected to a unit ramp function is

$$\theta_0(s) = \frac{500}{s^2(s + 2{\cdot}5 + j22{\cdot}2)(s + 2{\cdot}5 - j22{\cdot}2)}$$

As only an estimate of the transient performance is required we can content ourselves by quoting the damped natural frequency as 22·2 rad/s, and by measurement the damping ratio is found to be 0·1.

The steady-state error is

$$e(t)_{ss} = \lim_{s\to 0} s\,\frac{1}{s^2}\,\frac{1}{1 + 500/s(s+5)} = \frac{1}{100}$$

When the network is inserted $G(s)$ becomes

$$\frac{100(1 + 0{\cdot}1s)}{s(1 + 0{\cdot}2s)} = \frac{50(s+10)}{s(s+5)}$$

There is now a zero at $s = -10$ to add to the original plot (Fig. 5.11(*b*)).

The effect of the zero is to draw the original locus away from the positive half-plane, and now the real axis to the left of the zero is part of the locus. The asymptotes in this case are the real axis as

$$P - Z = 1 \qquad \alpha = \pm 180^\circ$$

The break-in and break-away points are found by differentiating $W(s)$:

$$W(s) = \frac{s(s+5)}{(s+10)}$$

The maxima or minima are determined when

$$\frac{dW(s)}{ds} = \frac{(s+10)(2s+5) - s(s+5)}{(s+10)^2} = 0$$

giving $s = -3$ and -17. The break-away point is $s = -3$, and there is a break-in point at $s = -17$. (It can be shown that the plot is circular, with centre $s = -10 + j0$ and radius $\sqrt{50}$.)

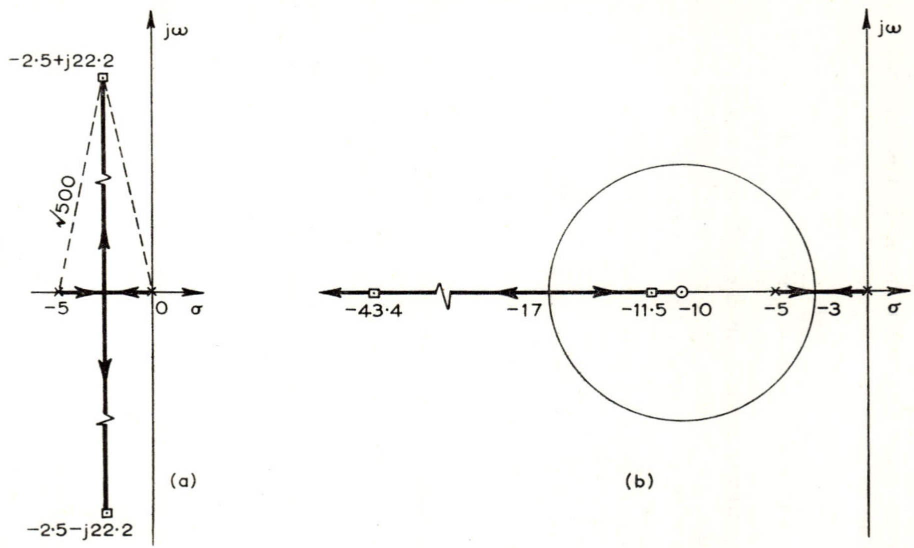

FIG 5.11

The closed-loop poles are obtained from the plot by trial and error (use could be made of a Spirule*) and are found to be at $-11{\cdot}5$ and $-43{\cdot}4$.

The transient response now has no oscillatory terms and is dominated by the closed-loop pole at $-11{\cdot}5$. For a unit ramp input the response will be similar to that of a first-order lag of time-constant $1/11{\cdot}5$ second:

$$\theta_0(s) = \frac{K}{s^2(s + 11{\cdot}5)}$$

The steady-state error is easily shown to be unchanged.

It is of interest to note that the break-in and break-away points can be found graphically by plotting a subsidiary root locus, noting

$$\frac{dW}{ds} = s^2 + 20s + 50 = 0$$

The roots of this equation can be found by writing it in the form

$$\frac{50}{s^2 + 20s} = -1$$

or

$$\frac{1}{s(s + 20)} = -1/K \quad \text{where } K = 50$$

Applying the normal rules for plotting root loci we have open-loop poles at 0 and -20, and the plot is that of the familiar second-order system. For $K = 50$,

$$|s(s + 20)| = 50$$

and for the closed-loop poles to be at $s = a$, $s = b$,

$$|(a)(20 - a)| = 50$$

giving $a = -17$ and -3 as before.

For a more complex system these points could be obtained by measurement from the subsidiary plot.

* A plastic device consisting of a circular protractor fitted with a rotatable arm. The arm is engraved with a logarithmic spiral to enable lengths to be multiplied together. The markings on the protractor enable the angles to be summed (Trade name of the Spirule Co., Whittier, California, U.S.A.).

Example 5.6

(*a*) Determine the characteristic function for the control system represented in Fig. 5.12, using block diagram algebra. What effect do the constants *A* and *B* have on the characteristic function?

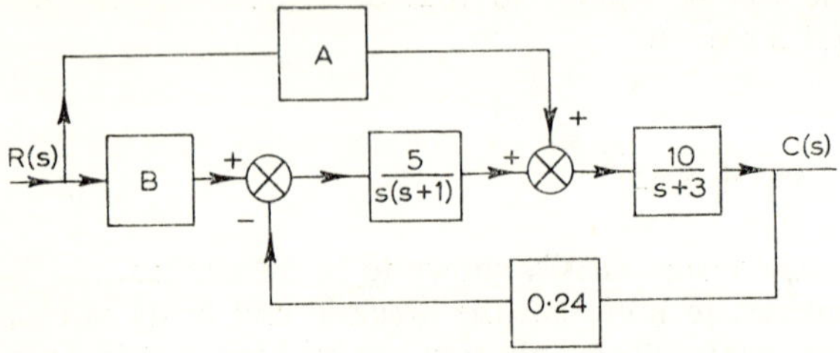

FIG 5.12

(*b*) Plot the poles of the characteristic function on the *s* plane and deduce whether the system is stable.

(*c*) Find a value for the steady-state loop gain which will result in a damping factor of approximately 0·3, and derive the time response of the system output to a unit-function input, with this value of loop gain. Take values for *A* and *B* of unity. (*Rugby*)

Solution

(*a*) The block diagram can be rearranged as in Fig. 5.13 (*a*) and can

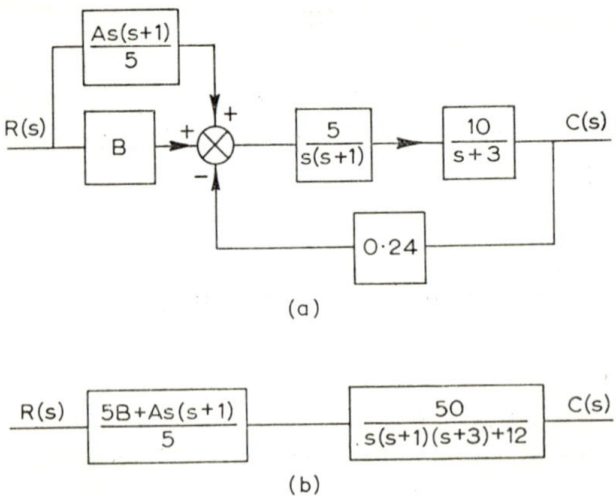

FIG 5.13

be further simplified as in (*b*). From this diagram it can be seen that the characteristic equation is

$$s(s + 1)(s + 3) + 12 = 0$$

Note that the constants A and B do not appear in this equation, the stability of the system being dependent only upon the closed-loop quantities.

(*b*) The poles of the characteristic function of Fig. 5.13 (*b*) can be found by plotting the root locus using the open-loop transfer function

$$G(s)\,H(s) = \frac{12}{s(s+1)(s+3)}$$

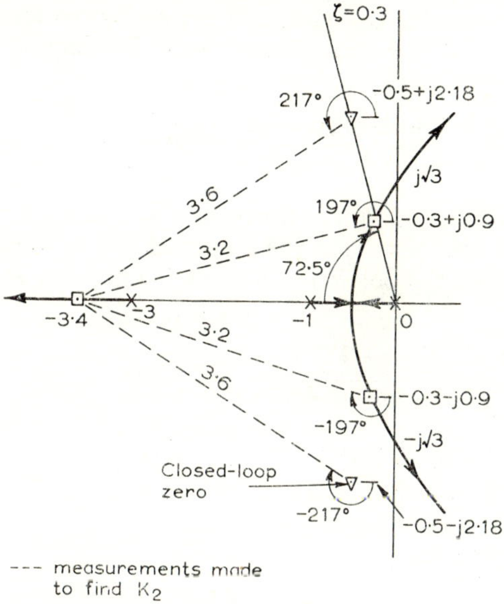

FIG 5.14

There are three poles; hence the asymptotes are at $\pm 60°$ and $180°$ and intersect the real axis at $s = -1{\cdot}33$. Routh's criterion gives the crossing points of the imaginary axis as $\pm j\sqrt{3}$ at a gain of 12. The system is at the limit of stability. The complete plot is shown in Fig. 5.14.

(*c*) On the root-locus plot of Fig. 5.14 a line has been drawn at $72{\cdot}5°$ to the real axis representing a damping ratio of 0·3. The value of open-loop gain where this intersects the root locus plot is 3·2. For this value of gain the third pole has moved along the real axis to $-3{\cdot}4$. Thus the complete transfer function of the system ($A = B = 1$) can be written

$$\left(\frac{s^2+s+5}{5}\right)\frac{50/s(s+1)(s+3)}{1+3{\cdot}2/s(s+1)(s+3)} = \frac{10(s^2+s+5)}{s(s+1)(s+3)+3{\cdot}2}$$

The poles of this expression have been obtained already, and it has zeros at $s = -0{\cdot}5 \pm j2{\cdot}18$. This has a time response to a unit step function input of the form

$$10[K_1 + K_2 e^{-3{\cdot}4t} + K_3 e^{(-0{\cdot}3+j0{\cdot}93)t} + K_3^{\star} e^{(-0{\cdot}3-j0{\cdot}93)t}]$$

where $K_3^{\star}$ is the complex conjugate of K_3. The constants K_1, K_2, etc., are the constants in the partial-fraction expansion of $C(s)$. They are obtained by multiplying $C(s)$ by a factor $(s + P_K)$, where P_K is the pole associated with the constant, and evaluating the resultant function when $s = P_K$ (this procedure only works for simple poles). To evaluate K_1:

$$K_1 = s\,\frac{(s + 0{\cdot}5 + j2{\cdot}18)(s + 0{\cdot}5 - j2{\cdot}18)}{s(s + 3{\cdot}4)(s + 0{\cdot}3 + j0{\cdot}93)(s + 0{\cdot}3 - j0{\cdot}93)}\bigg|_{s=0} = 1{\cdot}54$$

This however is the value of $C(s)$ at the point $s = 0$, assuming that the pole there does not exist. Thus the constants can be evaluated from measurements on the pole-zero diagram. To evaluate K_2 and K_3,

$$K_2 = \frac{(3{\cdot}6\underline{/217^\circ})(3{\cdot}6\underline{/-217^\circ})}{(3{\cdot}2\underline{/197^\circ})(3{\cdot}2\underline{/-197^\circ})(3{\cdot}4\underline{/180^\circ})} = -0{\cdot}37$$

$$K_3 = \frac{(1{\cdot}25\underline{/-81^\circ})(3{\cdot}1\underline{/86^\circ})}{(0{\cdot}95\underline{/108^\circ})(3{\cdot}2\underline{/16^\circ})(1{\cdot}8\underline{/90^\circ})} = 0{\cdot}71\underline{/-209^\circ}$$

and $K_3^{\star} = 0{\cdot}71\underline{/+209^\circ}$.

This gives the complete solution for $c(t)$ as

$$c(t) = 10[1{\cdot}58 - 0{\cdot}37e^{-3{\cdot}4t} + 1{\cdot}4e^{-0{\cdot}3t}\cos(0{\cdot}93t - 209^\circ)]$$

Example 5.7

The network shown in Fig. 5.15 is connected to the output terminals of a d.c. amplifier of gain K. Assuming the amplifier has infinite input impedance and zero output impedance, show that the forward transfer function of the combination is

$$G(s) = \frac{100Ks}{(s + 262)(s + 38)}$$

where $R_1C_1 = R_2C_2 = R_2C_1 = 10^{-2}$s.

If the output of the network is fed back to the amplifier input terminals, plot a root-locus diagram for the system, with (*a*) negative feedback, and (*b*) positive feedback. What value of gain K will just

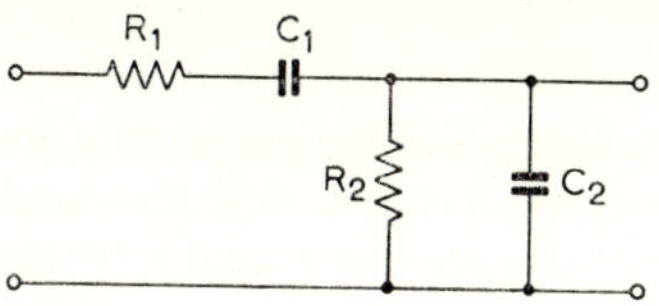

FIG 5.15

make the system oscillate? What value of K will cause continuous oscillations?

If a sinusoidal signal of amplitude 1 V at 50 Hz is fed into the positive feedback system, determine the amplitude and phase of the voltage at the network output terminals, when K has a value such that the damping ratio of the system is 0·5. (*Loughborough*)

Solution

The transfer function of the network and amplifier is given by

$$G(s) = \frac{KR_2C_1s}{(1 + sC_1R_1)(1 + sC_2R_2) + sC_1R_1}$$

Inserting the numerical value given for R_1C_1, R_2C_2 and R_2C_1,

$$G(s) = \frac{100Ks}{s^2 + 300s + 10\,000}$$

This is approximately equal to

$$G(s) = \frac{100Ks}{(s + 262)(s + 38)}$$

(*a*) For negative feedback the root locus is constructed using the standard rules and is shown in Fig. 5.16.

FIG 5.16

(*b*) For positive feedback the rules for plotting the root locus must be amended. The equation of the root locus can be seen by writing the closed-loop transfer function

$$\frac{C}{R}(s) = \frac{G(s)}{1 - G(s)}$$

(Note that the negative sign occurs because of the positive feedback.)

The poles of $C/R(s)$ are the roots of $G(s) = 1$, and this condition is satisfied by $|G(s) = 1|$, $\underline{/G(s)} = 0$. Hence the 0° and not the 180° contour must be plotted. In this particular case

$$\frac{C}{R}(s) = \frac{K's/(s+262)(s+38)}{1 - K's/(s+262)(s+38)}$$

$$= \frac{K's}{s^2 + (300 - K')s + (38 \times 262)}$$

where $K' = 100K$. This has poles at

$$s = \frac{-(300 - K') \pm \sqrt{[(300 - K')^2 - (4 \times 38 \times 262)]}}{2}$$

For $(300 - K')^2 \geqslant 4 \times 10^4$ the poles lie on the real axis.
For $(300 - K')^2 < 4 \times 10^4$ the poles lie at point (x, y) given by

$$x = -(300 - K')/2$$

$$y = \pm \frac{\sqrt{[4 \times 10^4 - (300 - K')^2]}}{2}$$

Note that

$$x^2 + y^2 = \frac{(300 - K')^2}{4} + \frac{(4 \times 10^4) - (300 - K')^2}{4} = 10^4$$

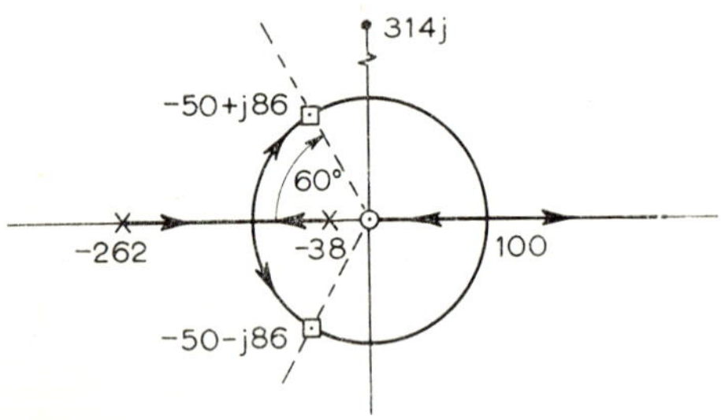

FIG 5.17

Hence the root locus for these values of K' is a circle with centre the origin and radius 100. The complete root locus is shown in Fig. 5.17.

The value of K for continuous oscillation is given when $x = 0$, i.e. $K' = 300$, $K = 3$. This can also be obtained by use of Routh's criterion.

A line representing $\zeta = 0{\cdot}5$ is drawn on the root-locus plot of Fig. 5.17. The intersection of this line and the root locus represents the position of a closed-loop pole for this value of damping ratio. Measurements at this point gives

$$K' = \frac{88 \times 227}{100} = 200$$

Hence the amplifier gain $K = 2$.

The gain and phase of the network can be obtained by evaluating the magnitude and phase angle at the point $s = j2\pi 50$ from the closed-loop pole-zero pattern.

Measurements made on Fig. 5.17 must be multiplied by 200 to give the magnitude of the response.

$$\left|\frac{C}{R}(j\omega)\right|_{\omega = 100\pi} = \frac{200 \times 314}{231 \times 402} = 0{\cdot}66$$

The corresponding phase shift is

$$90^\circ - 77^\circ - 82^\circ = -69^\circ \qquad \text{i.e. } 69^\circ \text{ lagging.}$$

This answer can be checked by substituting directly into the expression for $(C/R)(s)$ the values of K and $s = j100\pi$.

Example 5.8

The block diagram of a servomechanism with velocity feedback is shown in Fig. 5.18. Draw the root locus for this system where K is

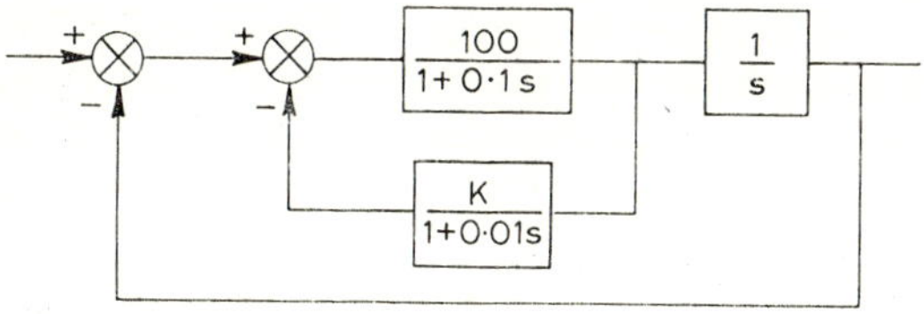

FIG 5.18

the variable parameter of the locus. Determine the value of K corresponding to a damped natural frequency of 20 Hz. (*Nottingham Univ.*)

Solution

The block diagram of Fig. 5.18 can be reduced to that of Fig. 5.19, which has unity feedback. The poles of the closed-loop transfer function are determined by the expression

$$\frac{1000(s+100)}{s[(s+10)(s+100)+K10^5]} = -1$$

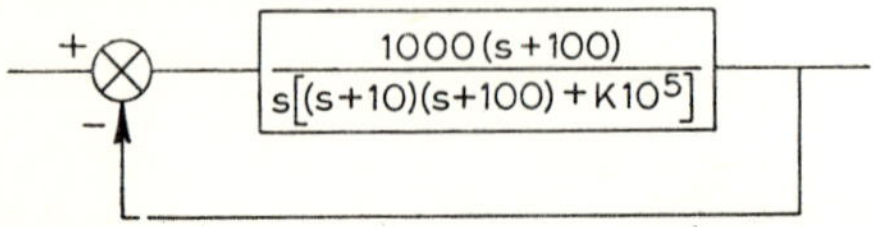

FIG 5.19

In this form the expression is not suitable for a direct root-locus plot, as K is the variable quantity. After rearrangement, however, it can be put into a form suitable for a standard root-locus plot:

$$\frac{5}{(s+100)(s^2+10s+1000)} = -\frac{1}{K10^5} = -\frac{1}{K'}$$

The plot as K' is varied as shown in Fig. 5.20. At $\omega = 2\pi 20 = 126$ rad/s, by measurement $K' = 1{\cdot}71 \times 10^4$; therefore $K = 0{\cdot}17$.

Example 5.9

The block diagram shown in Fig. 5.21 is for a closed-loop speed control system. Control is obtained by using a proportional-plus-integral controller, the transfer function of which is given by

$$G(s) = k\left(1 + \frac{1}{\tau_i s}\right) = \frac{k(s+a_i)}{s}$$

where $\tau_i = 1/a_i$ is the integral time-constant. The output member has an inertia J and is driven by a motor whose transfer function approximates to a simple lag and gain factor k_m. T_m and T_L are the motor torque and load torque respectively. Show that the closed-loop transfer function is

$$C(s) = \frac{K(s+a_i)R(s) - 2{\cdot}95(s+\sigma_m)T_L(s)}{(s+q_1)(s+q_2)(s+q_3)}$$

where $K = k(K_m/J) = 10k$, and q_1, q_2, q_3 are roots of the characteristic equation.

Fig. 5.22 shows root loci for the system for the following values of a_i; 0, 1, 4/3, 2, 4 and 6 with lines of constant gain superimposed on

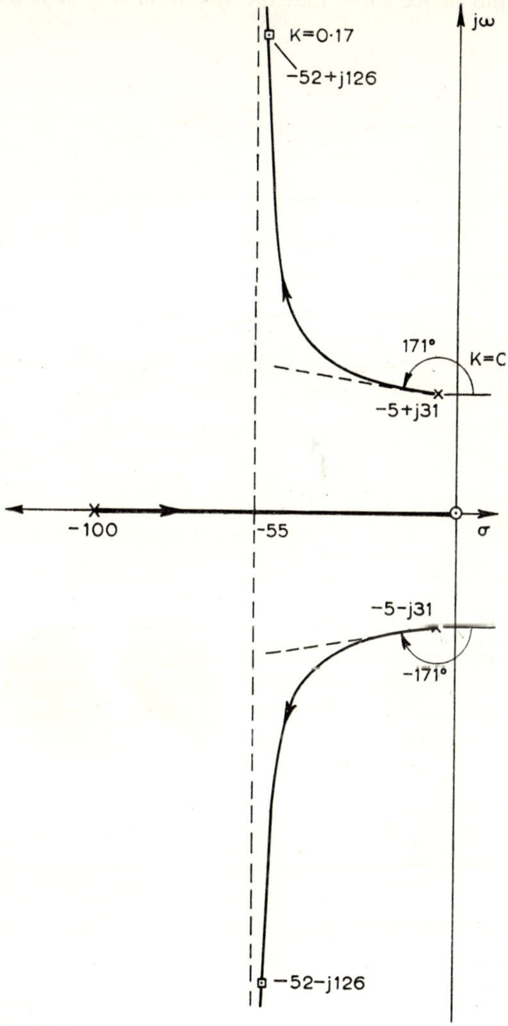

FIG 5.20

the loci. By inspection find the range of values of a for which the system is inherently stable. Use the diagram also to complete the following table, and hence show that the system is very sensitive to

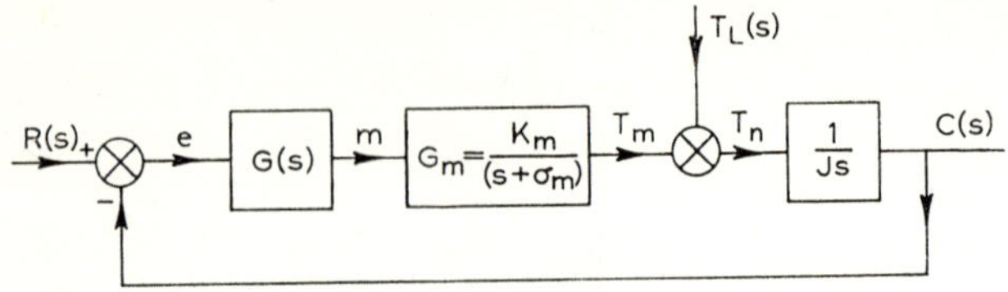

FIG 5.21

$K_m = 3{\cdot}4$ N per unit of m $\quad \sigma_m = 12\,s^{-1}$
$J = 0{\cdot}34$ kg-m^2 $\quad \tau_m = 0{\cdot}0833$ s

the value of the integral time-constant. Determine the unit step response of the system for condition (c) given in the table, assuming there is no load torque disturbance.

Condition	K	τ_i (second)	Roots of characteristic equation	t_p, time to peak (second)	M_0 per unit maximum overshoot
a	48	1			
b	48	$\frac{3}{4}$			
c	48	$\frac{1}{2}$			

(*Loughborough**)

Solution

If $T_L(s)$ is transferred to the input it becomes

$$T_L(s) = \frac{(s+\sigma_m)}{K_m}\frac{1}{G(s)}$$

The expression for $C(s)$ becomes

$$C(s) = \frac{K_m G(s)/Js(s+\sigma_m)}{1 + K_m G(s)/(s+\sigma_m)Js}\left[R(s) - \frac{(s+\sigma_m)T_L(s)}{K_m G(s)}\right]$$

$$= \frac{K_m G(s) R(s) - (s+\sigma_m)T_L(s)}{Js(s+\sigma_m) + K_m G(s)}$$

Putting $G(s) = k(s + a_i)/s$,

$$C(s) = \frac{K_m k(s+a_i)R(s)/s - (s+\sigma_m)T_L(s)}{Js(s+\sigma_m) + K_m k(s+a_i)/s}$$

Writing $K = K_m k/J$ and substituting the value of J,

$$C(s) = \frac{K(s + a_i)R(s)/s - 2{\cdot}95(s + \sigma_m)T_L(s)}{s(s + \sigma_m) + K(s + a_i)/s}$$

$$= \frac{K(s + a_i)R(s) - 2{\cdot}95s(s + \sigma_m)T_L(s)}{(s + q_1)(s + q_2)(s + q_3)}$$

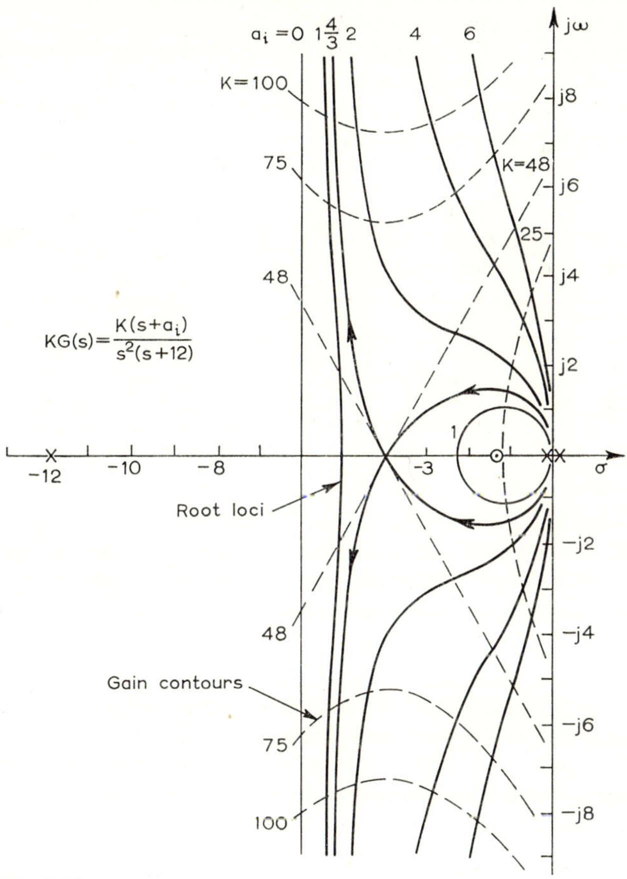

FIG 5.22

The open-loop transfer function is $K(s + a_i)/s^2(s + 12)$, and in Fig. 5.22 the root loci corresponding to several values of a_i are drawn. The system is inherently stable for all values of a_i since the root loci do not cross the imaginary axis.

The values of a_i can be easily calculated as $a_i = 1/\tau_i$; hence

$a_i = 1, 4/3$ and 2 for values of τ of 1, $\frac{3}{4}$ and $\frac{1}{2}$ respectively. The table can now be completed for each of the values of a_i.

(1) $a_i = 1$

With $K = 48$, the characteristic equation becomes

$$s^3 + 12s^2 + 48s + 48 = 0 \qquad (1)$$

The roots of this equation (equivalent to the closed-loop poles) can be obtained from the root locus plot.

From Fig. 5.22 the complex roots are determined by the intersection of the gain contours and the root loci for $a_i = 1$. When $K = 48$ the complex roots are approximately $s = -5{\cdot}2 \pm j1{\cdot}9$, and factorizing eqn. (1) gives the real root as approximately $s = -1{\cdot}6$. The positions of these poles on the complex plane are shown in Fig. 5.23 (*a*).

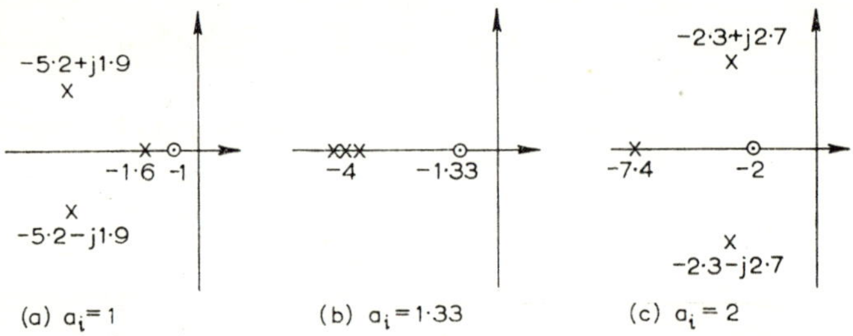

FIG 5.23

There is no simple expression for rise time and peak overshoot for systems having complicated pole-zero patterns. Simplification is possible, however, if the system response is governed by a pair of dominant complex poles. Poles can be considered dominant if:

(i) All other poles are far enough to the left of the axis so that the transients associated with them die away more rapidly than that due to the dominant poles.
(ii) The poles are not far enough to the left for condition (i) to apply but are close to a zero so that the weighting given to the transients due to such poles is small.

If a pair of poles can be considered dominant, the transient terms due to any other poles can be neglected. However, the effect of weighting of the dominant transient due to these other poles cannot be neglected, and here the formula for rise time t_p of eqn. (3.9) is modified as follows:

$$t_p = \frac{1}{\omega_d}\left[\frac{\pi}{2} - \underline{/Z(s)} + \underline{/P(s)}\right]$$

where $\angle Z(s)$ is the sum of the angles of the zeros of $\frac{C}{R}(s)$ to a dominant pole, and $\angle P(s)$ is the sum of the angles from all other poles of $\frac{C}{R}(s)$ to the dominant pole, including the angle from the conjugate pole.

(Note the above equation reduces to eqn. (3.9), if we have only a pair of dominant complex poles, since $\angle Z(s) = 0$ and $\angle P(s) = \pi/2$.)

A simplified formula for peak overshoot can be obtained for a unity feedback system that is type 1 or higher:

$$M_p = \frac{\Pi_1 \, \Pi_2}{\Pi_3 \, \Pi_4} e^{\sigma t_p}$$

where

Π_1 = product of the distances of all poles other than the dominant poles to the origin
Π_3 = product of the distances of all poles to a dominant pole excluding the distance from the conjugate pole
Π_2 = product of the distances from all zeros to a dominant pole
Π_4 = product of the distances from all zeros to the origin
σ = real part of the location of the dominant poles

Use of the formulae here gives

$$t_p = (1/1{\cdot}9)[90^\circ - 155^\circ + (151^\circ \mid 90^\circ)](\pi/180) - 1{\cdot}6\,\mathrm{s}$$

$$M_p = \frac{1{\cdot}6 \times 4{\cdot}6}{3{\cdot}8 \times 1} e^{-5{\cdot}2 \times 1{\cdot}6} \text{ (negligible)}$$

(2) $a_i = 1{\cdot}33$

When $a = 1{\cdot}33$ there is a triple root of eqn. (1) at $s = -4$ (see Fig. 5.23 (*b*)). There is no overshoot, $M_0 = 0$ and $t_p = \infty$, i.e. it is a very sluggish system.

(3) $a_i = 2$

In this case the complex poles are at $s = -2{\cdot}3 \pm j2{\cdot}7$, and the real pole is found to be at $s = -7{\cdot}4$. These are shown in Fig. 5.23 (*c*). For this condition

$$t_p = (1/2{\cdot}7)[90^\circ - 97^\circ + (90^\circ + 28^\circ)](\pi/180) = 0{\cdot}7\,\mathrm{s}$$

$$M_p = \frac{7{\cdot}4 \times 2{\cdot}7}{5{\cdot}7 \times 2} e^{-0{\cdot}7 \times 2{\cdot}3} = 0{\cdot}35\,\text{p.u.}$$

The table in completed form is

Condition	K	τ_i (second)	Roots of characteristic equation	t_p time to peak (second)	M_p per unit maximum overshoot
a	48	1	$-5 \pm j2$	1·7	negligible
b	48	$\frac{3}{4}$	-4 (triple root)	no overshoot	
c	48	$\frac{1}{2}$	$-2{\cdot}3 \pm j2{\cdot}7$	0·7	0·35

Note. For the condition where $a_i = 1$ the system could be approximated by considering that the pole and zero on the real axis cancel, and the system can be considered as second order having the characteristic equation

$$(s + 5 - j2)(s + 5 + j2) = s^2 + 10s + 29 = 0$$

The complex poles have a damping ratio of 0·93, giving rise to a negligible overshoot. (Use could be made of eqns. (3.9) and (3.8) to calculate rise time and overshoot.)

When $a_i = 1{\cdot}33$, only real poles would exist and the system would have a sluggish response. On the other hand, when $a_i = 2$ the zero has a large effect on the response and cannot be neglected.

If the response were considered due to the complex poles alone this would give a damping ratio of 0·65, indicating an overshoot of the order 0·07 p.u. The actual overshoot taking the zero into account is 0·35 p.u., showing that the system is very sensitive to the integral time-constant.

Example 5.10

The control system shown in Fig. 5.24 has an inner loop whose gain K_m is adjusted to give a critically damped response of the output y to a step input x. The amplifier gain K_a and the feedback time constant τ_f can be varied, but all other parameters are fixed due to the inherent characteristics of the components used. Sketch the root-locus diagram for the system as the overall forward gain is varied from 0 to ∞, and justify the break-away and instability points.

From the root-locus diagram, determine:

(i) The range of gain for which the system has two oscillatory terms of different frequency and amplitude in the output response to a step input.

(ii) Whether or not the system can be stable when there is only one oscillatory term in the response.
(iii) The limiting value of τ_f to ensure that there is only one oscillatory term in the response.
(iv) The amplitude, frequency and phase shift of the higher-frequency oscillatory term in the step response, when the closed-loop damping ratio ζ_{CL} for this term is 0·25, and find the value of K_a at this value of ζ_{CL}.

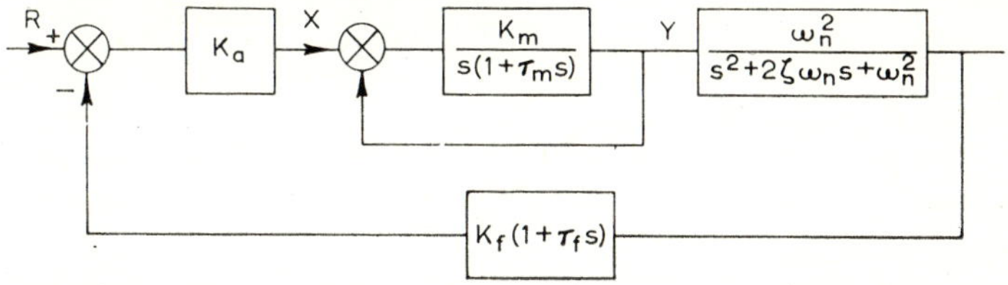

FIG 5.24

$\tau_m = 0{\cdot}05\,\text{s}$ $\quad K_f = 1$
$\omega_n = 100\,\text{s}^{-1}$ $\quad \tau_f = 0{\cdot}025\,\text{s}$
$\zeta = 0{\cdot}3$

Illustrate the change of configuration of closed-loop poles caused by inserting phase-advance, having a transfer function of the form $\alpha(1 + \tau s)/(1 + \alpha\tau s)$ in the amplifier stage and replacing the feedback transfer function by unity feedback. For this illustration assume $\alpha = 0{\cdot}1$ and $\tau = 0{\cdot}025$ s. (*Loughborough*)

Solution

A root-locus plot can be first constructed for the portion of the block diagram from X to Y. The open-loop transfer function of this portion is given by

$$\frac{K_m}{s(1 + 0{\cdot}05s)} = \frac{20K_m}{s(20 + s)}$$

The root locus for an expression of this form, having two poles on the real axis, has been dealt with in the introduction.

The value of K_m for critical damping is at the break-away points from the real axis. The break-away condition is

$$\frac{dW(s)}{ds} = 0$$

and substituting $W = s(s + 20)$ gives $s = -10$; hence $K_m = 5$.

The transfer function Y/X is now given by

$$\frac{Y}{X} = \frac{100}{(s + 10)^2}$$

and the open-loop transfer function for the main loop becomes

$$G(s)\,H(s) = \frac{100K_a}{(s+10)^2}\,\frac{\omega_n^2}{s^2 + 2\zeta\omega_n s + \omega_n^2}\,K_f(1 + \tau_f s)$$

Substituting the values given for K_f, ω_n and τ_f, this becomes

$$G(s)\,H(s) = \frac{10^6 K_a\,0{\cdot}025(s+40)}{(s+10)^2(s^2+60s+10^4)}$$

$$= \frac{K'(s+40)}{(s+10)^2(s+30+j95)(s+30-j95)}$$

where $K' = 2{\cdot}5 \times 10^4 K_a$.

The pole-zero pattern, together with the root-locus plot as K' is varied from 0 to ∞, is shown in Fig. 5.25. The characteristic equation is

$$s^4 + 80s^3 + 11\,300s^2 + [(206 \times 10^3) + K']s + 10^6 + 40K' = 0$$

and application of Routh's criterion gives the imaginary axis crossing points as $s = \pm j94{\cdot}3$ when $K' = 5{\cdot}07 \times 10^{-5}$.

The break-in point to the real axis can be verified as follows:

$$W = \frac{(s+10)^2(s^2+60s+10^4)}{s+40}$$

$$\frac{dW}{ds} = 0 \text{ at a break-in point}$$

$$0 = \frac{(s+40)[2(s+10)(s^2+60s+10^4) + (s+10)^2(2s+60)] - (s+10)^2(s^2+60s+10^4)}{(s+40)^2}$$

which simplifies to

$$\frac{(s+10)(s+30)(s+40)}{(s^2+60s+10^4)(s+70)} = -\frac{1}{2}$$

The root locus of this expression is given in Fig. 5.26 (the branches from the complex poles have not been shown). In the region $s = -40 \rightarrow -70$ the function has magnitude 0·5 at $s = -62{\cdot}4$, and this is the required break-in point. Hence from Fig. 5.25 we can state:

(i) The system has two oscillatory modes as K' varies from 0 to $1{\cdot}23 \times 10^6$. (K_a from 0 to 50).

(ii) $K' = 5 \times 10^5$; $K_a = 20$ when the system goes unstable; therefore it cannot be stable with only one oscillatory mode.

(iii) The limit of τ_f is when the zero corresponds with one of the poles at $s = -10$; i.e. $\tau_f = 0{\cdot}1$ s (see Fig. 5.27).

(iv) For $\zeta_{CL} = 0{\cdot}25$, $K' = 104\,000$ from the root locus. Hence $K_a = 4$, and frequency $= 93\,\text{rad/s} = 14{\cdot}8\,\text{Hz}$.

The phase-advance network has a transfer function

$$\frac{0{\cdot}1(1 + 0{\cdot}025s)}{(1 + 0{\cdot}0025s)} = \frac{s + 40}{s + 400}$$

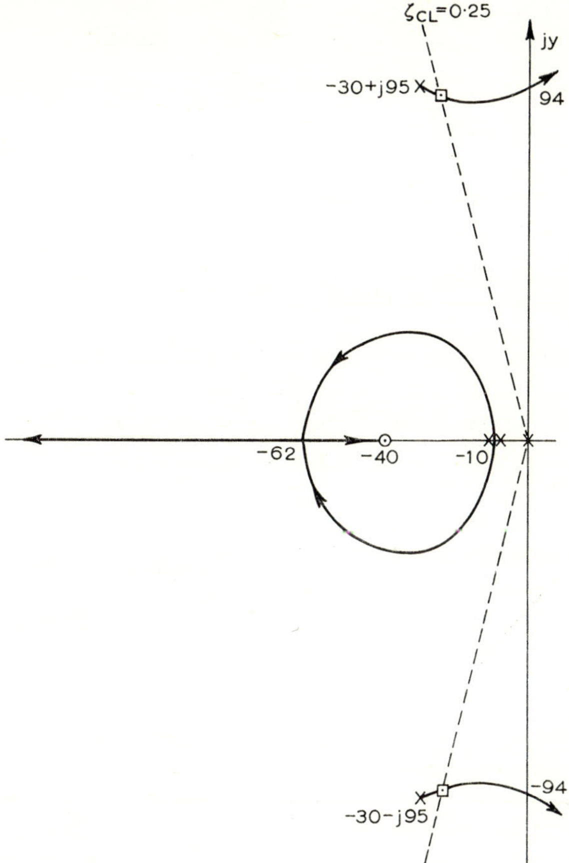

FIG 5.25

The new open-loop transfer function becomes

$$\frac{K_a 10^6(s + 40)}{(s + 10)^2(s^2 + 60s + 10^4)(s + 400)}$$

$$= \frac{K'(s + 40)}{(s + 10)^2(s^2 + 60s + 10^4)(s + 400)}$$

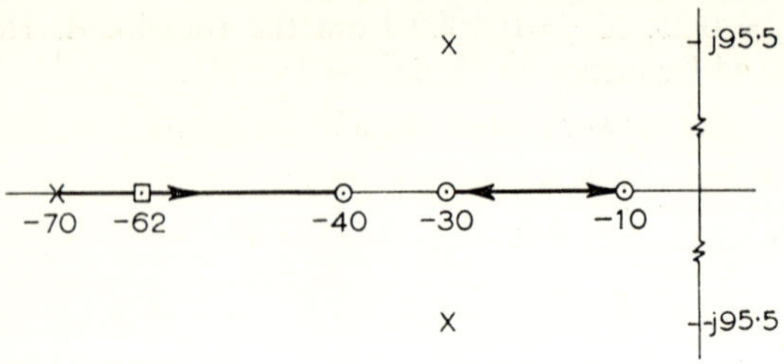

Fig 5.26

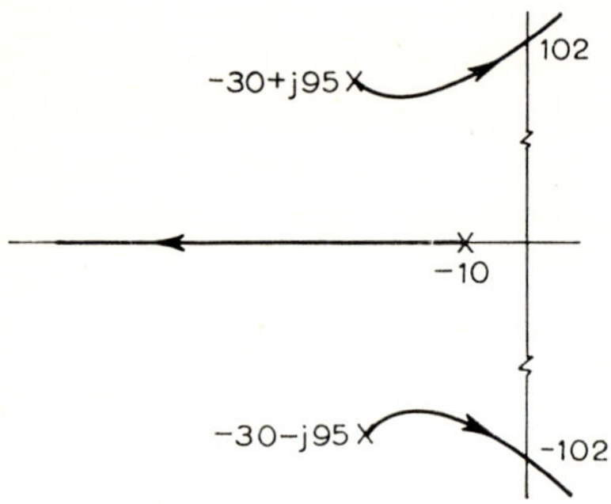

Fig 5.27

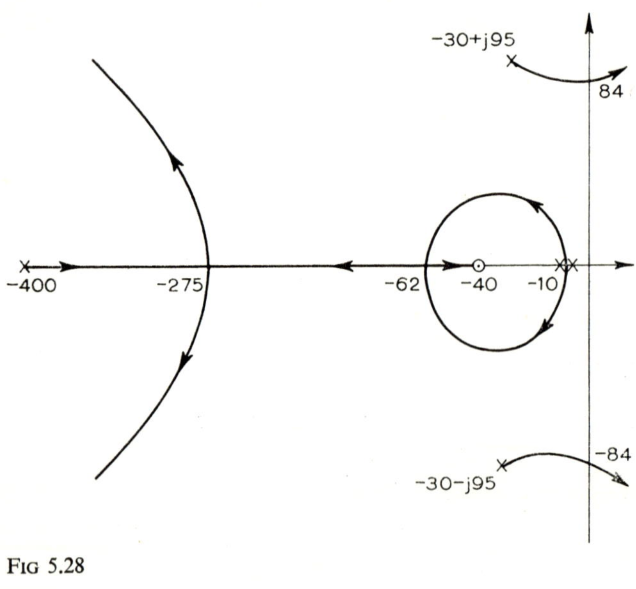

Fig 5.28

The root-locus plot for this open-loop transfer function is shown in Fig. 5.28.

Example 5.11

Explain, with reference to the Nyquist diagram, the effect of a finite time lag in the forward path of a control system.

Fig. 5.29 shows a machine which is used to reduce material to a desired thickness. A cutting wheel C is raised and lowered under the

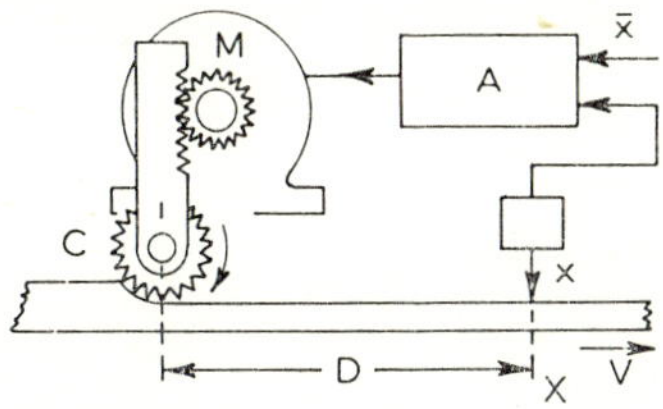

FIG 5.29

action of the control motor M so as to cut the material as it passes continuously from left to right at speed V. The thickness x after cutting is measured automatically at the point X, and the difference between this value and a desired value $\bar{x}$ is used to control the motor M via amplifier A. The transfer function relating vertical movement of C with the error $(\bar{x} - x)$ is

$$\frac{K}{j\omega(1 + j\omega\tau)}$$

where $K = 100\,\text{s}^{-1}$, $\tau = 0{\cdot}01$ s.

If $V = 1{\cdot}52$ m/s, find the maximum permissible distance D before sustained oscillations are produced. (*Brunel Univ.*)*

Solution

The problem is an example of a system having a transport lag or dead time. The transfer function of the transport lag element is e^{-Ts}. It is of interest to investigate this system by using the root-locus approach. (The reader should also attempt the solution by a frequency-response method.)

The open-loop transfer function in this example is

$$\frac{1\,000\,e^{-Ts}}{s(s + 10)}$$

The system with negligible dead time would have the second-order loci (Fig. 5.30).

$e^{-j\omega\tau} =$

The term $e^{-s\tau}$ at point $s = \sigma + j\omega$ becomes $e^{-\sigma\tau}e^{-j\omega\tau}$. The magnitude scaling on the root locus alters by a factor $e^{-\sigma\tau}$, and there is additional phase shift of $\omega\tau$ radians. This gives a root locus as shown dotted in Fig. 5.30. When this crosses the imaginary axis $\sigma = 0$, and

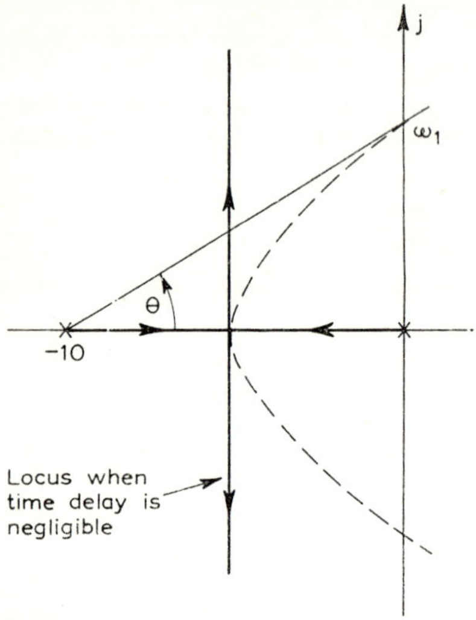

FIG 5.30

the magnitude at this point can be found by standard root-locus techniques. The magnitude is known to be 1 000 and this can be used to obtain the crossing point $j\omega_1$:

$$\omega_1\sqrt{(10^2 + \omega_1^2)} = 1\,000$$

whence $\omega_1 = 30{\cdot}8$ rad/s.

The angle at this point neglecting the lag term would be $-90° - \theta$, where $\theta = \tan^{-1} \omega_1/10 = 72°$. Hence the total phase shift is $-162°$.

As the root locus now passes through the point $s = j\omega_1$, the term $e^{-j\omega_1\tau}$ must give an extra 18° lag:

$$\omega_1\tau = 18\pi/180$$

giving $\tau_1 = 10{\cdot}4$ ms. At $1{\cdot}52$ m/s this is equivalent to a distance of $1{\cdot}58$ cm.

PROBLEMS

5.1 Discuss methods of assessing the closed-loop transient response of a control system when the open-loop transfer function is known.

If an open-loop transfer function is given by

$$KG(s) = \frac{K}{(1 + s)(1 + 0{\cdot}1s)(1 + 0{\cdot}05s)}$$

estimate the maximum value of loop gain K for which the transient response of the system when the loop is closed will be adequately damped.

Explain how the system could be modified so that the gain K could be increased while maintaining a satisfactory response. (*Cambridge Univ.*)

(*Ans.*) For $\zeta = 0{\cdot}5$, $K = 5{\cdot}5$.

5.2 A feedback control system has an open-loop transfer function

$$KG(s) = \frac{K}{s(1 + s)(1 + 0{\cdot}5s)}$$

Determine, using a root-locus plot, the open-loop gain K required to obtain a closed-loop step response corresponding to a damping ratio of 0·6 for the complex roots, and estimate the damped frequency of oscillation of the step response.

Determine also the restriction on the open-loop gain which ensures stability, and specify the form of the closed-loop step response. (*Cambridge Univ.*)

Ans. $K = 0{\cdot}4$, $\omega_d = 0{\cdot}5$ rad/s; $K = 3$.

5.3 A servomechanism has unity feedback and the following forward-loop transfer function

$$\frac{K(1 + 0{\cdot}5s)}{s(1 + 0{\cdot}1s)(1 + 0{\cdot}05s)}$$

Plot the root locus; determine the value of K corresponding to a damped natural frequency of 0·4 Hz. What is the damping ratio? (*Nottingham Univ.*)

Ans. $K = 5$, $\zeta = 0{\cdot}4$.

5.4 Indicate a graphical method for determining the gain and phase shift of a network having a transfer function

$$\frac{s + \alpha}{(s + b)(s + \alpha + j\beta)(s + \alpha - j\beta)}$$

to a sinusoidal signal of angular frequency ω rad/s. (*Manchester Univ.*)

5.5 In the block diagram of Fig. 5.31,

$$G(s) = \frac{1}{s(s+2)(s+6)} \quad \text{and} \quad H(s) = 1$$

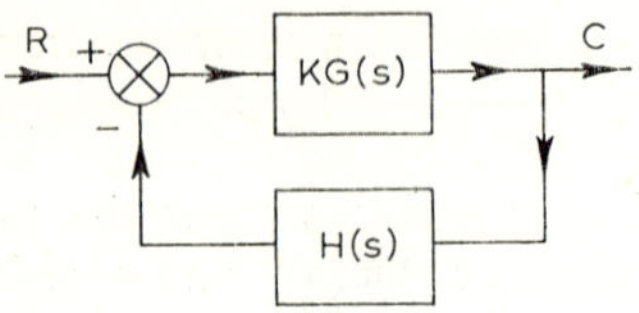

FIG 5.31

Sketch the root-locus plot of this system and determine the value of K required to yield a damping ratio of 0·5. (*Liverpool Univ.*)

Ans. 15.

5.6 Describe the root-locus method for assessing the performance of control systems and discuss the rules that facilitate rapid construction of the locus. Show how you would use the root-locus plot to evaluate the gain of a system for a specific value of damping ratio. (*Liverpool Univ.*)

5.7 A control system has a forward transfer function

$$KG(s) = \frac{K}{s(1 + 0{\cdot}2s)(1 + 0{\cdot}1s)}$$

and a unity feedback transfer function. Plot the root locus and from it evaluate the limiting value of K for stability. (*Liverpool Univ.*)

Ans. 15.

5.8 The open-loop transfer function of a control system is

$$G(s) = \frac{K(s+4)}{s(s^2 + 6s + 13)}$$

Observation of transient responses of this system suggested the existence of a time-constant of 2 seconds in the system closed-loop transfer function in addition to an oscillatory mode.

On the basis of this information, and using the root-locus technique, determine the closed-loop transfer function and from it, the time response to a unit step input.

Note. Reasonable approximations in the root locus will be accepted. (*Birmingham Univ.*)

Ans. $\dfrac{1{\cdot}45(s+4)}{(s+0{\cdot}5)(s^2+5{\cdot}6s+12)}$; $1{\cdot}0 - 1{\cdot}1\,e^{-0{\cdot}5t} + 0{\cdot}17\,e^{-2{\cdot}8t}\sin(2t - 3{\cdot}7)$

5.9 Explain the terms "Nyquist plot," "Bode plot" and "root locus." Describe how you might set about designing a speed control system when the load, gearbox and motor to drive it are already decided and fixed parts of the system.

Explain with reference to a load and motor combination with the transference

$$P(s) = \frac{20}{s(s + 0{\cdot}1)(s + 10)}$$

how you ensure (*a*) stability and (*b*) a settling time of less than 2 seconds. (*Swansea*)

5.10 What information does the root-locus technique offer to the designer about the behaviour of a feedback control system?

Sketch the root locus of a system with a loop transfer function

$$G(s) = \frac{K}{s(s + 3)(s^2 + 2s + 2)}$$

and briefly describe the procedure.

In this same system, it is desirable to control the damping so that the oscillatory modes of response diminish in amplitude in one cycle to not less than one-tenth of the preceding value. What is the limiting value of K to satisfy this specification? (*Swansea*)

Ans. 3·5, $\zeta = 0{\cdot}34$.

5.11 Describe how the root-locus diagram can be used to indicate the stability boundaries of a control system. Indicate how one can calculate the performance of a system for a given value of the gain.

Sketch the root locus diagram for a control system whose open-loop transfer function is given by

$$s^2\theta_0 = \omega_0{}^2(1 + \tau_1 s)\theta_e$$

Given that $\tau_1 = 4$ seconds, find the value of the system gain so that both roots are just critically damped. (*Southampton Univ.*)

Ans. $\omega_0 = 0{\cdot}5$.

5.12 A unity-feedback type-one system has two exponential lags in the forward path of time-constants 0·1 second and 0·05 second respectively. Use the root locus method to establish the value of K, the velocity error constant for which the relative damping ζ of the oscillatory mode is 0·45.

State the value of the transient angular frequency of oscillation and show the location of the three closed-loop roots on the root-locus diagram. Sketch the sections of the root locus. (*I.E.R.E.*)

Ans. $K = 5{\cdot}9s^{-1}$, $\omega_d = 6{\cdot}2$ rad/s.

5.13 (*a*) Discuss the value of the root-locus technique for feedback control system analysis.

(*b*) Outline the salient steps in plotting a root-locus diagram, taking as an example the open-loop transfer function

$$KG(s) = \frac{K(1 + 0{\cdot}33s)}{s(s + 4)(s + 6)(s^2 + 2s + 2)}$$

Sketch approximately the locus for the system without obtaining accurate solutions to the break-away point, etc. The angle of departure from the complex poles is $-50°$ and $50°$.

(*c*) Explain the effects on a root locus of adding (i) open-loop poles, and (ii) open-loop zeros, and hence show how compensation techniques can be displayed on the root-locus. (*Brunel Univ.*)

5.14 A direct-coupled amplifier has the following frequency response

$$A = \frac{-A_0}{(1 + j\omega/\omega_0)(1 + j\omega/2\omega_0)(1 + j\omega/4\omega_0)}$$

where $\omega_0 = 10\,\text{rad/s}$.

Plot with reasonable accuracy the root-locus of the feedback system formed of the above amplifier and a feedback network for which $\beta = 0{\cdot}125$. A_0 is the only parameter which can be varied.

Additional components are now added to the feedback network so that β becomes

$$0{\cdot}125\,\frac{1 + j\omega/0{\cdot}9\omega_0}{1 + j\omega/8\omega_0}$$

Sketch the modified root-locus and comment on the changes in the feedback system behaviour for various A_0. (*Southampton Univ.*)

5.15 In the amplifier circuit shown in outline in Fig. 5.32, the first and third transistors operate as virtual-earth feedback stages, while the second has emitter-resistance feedback. Each stage may be taken as controlled by its feedback and independent of the other stages. The β for the third transistor is 50. Non-ideal behaviour of the output transformer is allowed for by the inductance and capacitance shown, and overall negative feedback is obtained when the input switch is at position 2.

Obtain the Thévenin equivalent for the circuit between X and earth, to the left of X, when the switch is in position 1, and hence evaluate the coefficients *a*, *b* and *c* in the expression for the open-loop transfer function:

$$\frac{V_f(s)}{V(s)} = \frac{-10}{(1 + as)(1 + bs + cs^2)}$$

Plot the poles of this transfer function in the s plane and hence estimate the phase angle between V_f and V at the frequency for which they are equal in magnitude. (*Cambridge Univ.*)

Ans. $a = 20 \times 10^{-6}$, $b = 10^{-6}$, $c = 10^{-12}$, $-310°$.

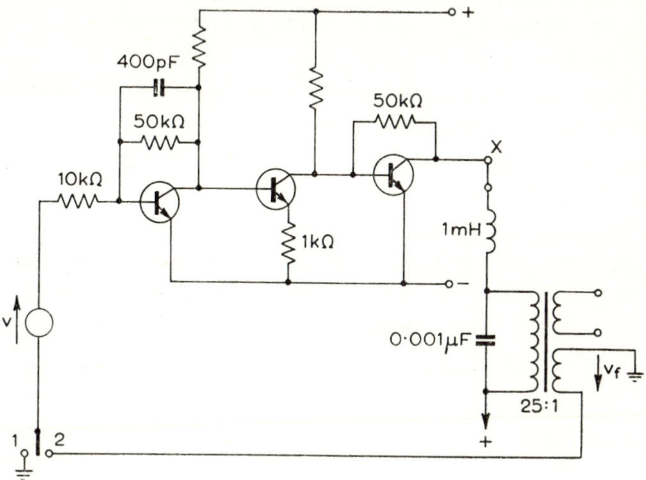

FIG 5.32

5.16 A second-order system has the characteristic equation

$$s^2\theta_0 + 2\zeta\omega_n s\theta_0 + \omega_n{}^2\theta_0 = \theta_i$$

Show that the locus of the roots of the auxiliary equation are circles for ω_n constant and straight lines for ζ constant. Describe how this general knowledge of the location of the roots of the characteristic equation may be used in interpreting a root-locus plot for a complex system.

Sketch the root-locus plot for a system having the open-loop transfer function

$$G(s)\,H(s) = \frac{K(s+1)}{(s+4)(s+0{\cdot}8+j0{\cdot}1)(s+0{\cdot}8-j0{\cdot}1)}$$

Estimate the value of K required to give a performance having a similar response characteristic to a second-order system with damping ratio $\zeta = 0{\cdot}6$. (*Lanchester*)

Ans. $K = 12$.

5.17 A control system has an open-loop transfer function

$$G(s)\,H(s) = K/s(s+3)(s+5)$$

Sketch the root locus for this system, and using this locus, determine the values of K which ensures that the close-loop control system

(*a*) has a critically damped transient response;
(*b*) has an adequately damped transient response;
(*c*) is about to go unstable.

Plot the pole-zero configuration for each value of K and estimate the time for the transients to decay in case (*b*).

Ans. (*a*) $K = 8$, (*b*) $K = 18$, $\zeta = 0{\cdot}6$, (*c*) $K = 120$; settling time (2% of the final value) = 4 s.

5.18 How can information about the oscillating part of the transient response of a given system be obtained from the root-locus diagram?

The transfer function in the system of Fig. 5.33 is given by

$$KG(s) = \frac{K(1+0{\cdot}5s)}{s(1+s)(1+0{\cdot}25s)}$$

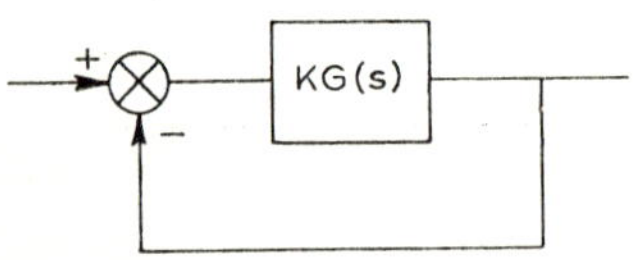

FIG 5.33

Sketch the root-locus diagram for the closed-loop system.

The value of K which will give the oscillating part of the response a damping ratio of $1/\sqrt{2}$ is to be determined. Show that the determination of K involves the solution of the equation

$$\tan^{-1} x/(1-x) + \tan^{-1} x/(4-x) - \tan^{-1} x/(2-x) = 45°$$

where $-x$ is the real part of the complex roots. Given that the solution of the latter equation is $x = 0{\cdot}62$, determine the required value of K. (*Edinburgh Univ.*)

Ans. $K = 0{\cdot}7$.

5.19 The forward-path transfer function of a unity feedback system is

$$G(s) = \frac{K(s+2{\cdot}5)}{s(s+2)(s+6)(s+15)}$$

Plot the root locus in the region of damping ratio = 0·4 (and sketch it elsewhere) using a plot size of 10 units = 1 Spirule unit. Determine the loop sensitivity which results in complex dominant roots of the above damping ratio. Determine and mark on your plot the positions of the other two roots for the same value of loop sensitivity and write down the transfer function of the closed-loop system.

Hence calculate the magnitude and damping factor of the slower exponential decay term in the time response to a unit step input.

Describe and illustrate in general terms the influence of the factors $(s + 2{\cdot}5)/(s + 15)$ on the simpler but less effective system

$$G(s) = \frac{K}{s(s + 2)(s + 6)}$$

(*Nottingham Univ.*)

Ans. $\dfrac{358(s + 2{\cdot}5)}{(s + 2{\cdot}7)(s + 16{\cdot}8)(s^2 + 3{\cdot}6s + 20)}$, $-0{\cdot}35\,e^{-2{\cdot}7t}$

5.20 Fig. 5.34 is the block diagram for a process-control system where c is the controlled variable, r is the reference variable, and K is the gain factor of an electronic amplifier.

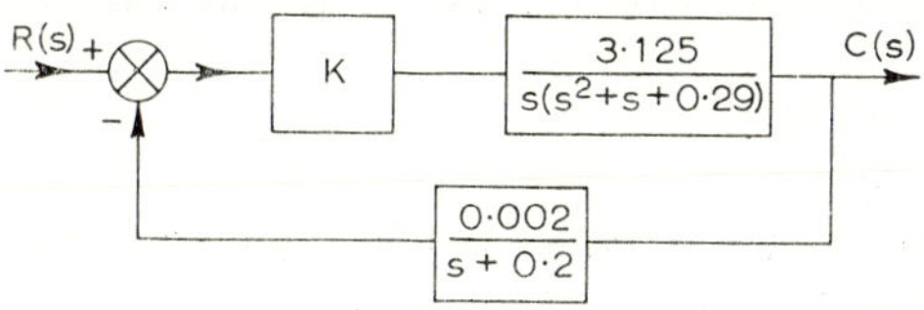

FIG 5.34

Sketch an approximate root locus for the system and hence estimate the value for K such that the system damping ratio is 0·4. Show that the resultant ramp error coefficient K_1 and the damped natural frequency for the system are approximately 11 and 0·125 rad/s respectively.

It is required to speed up the response of the system and furthermore to improve K_1. Design a suitable resistor-capacitor cascade compensator to achieve this aim and determine, by drawing a root locus plot, the new values for K_1 and ω_d for a system damping ratio of 0·4. Comment on the acceptability of all roots of the system.

Capacitor values should not exceed 50 μF and the value for K, due to drift effects, should not exceed 500. (*Nottingham Univ.*)

Ans. $K = 1{\cdot}1$.

Chapter 6

Nyquist and Inverse Nyquist Diagrams

The Nyquist diagram was used to determine the absolute stability of a system in Chapter 4. In practice it is of the utmost importance to obtain a measure of the degree of stability of a system. The Nyquist diagram can be used for this purpose, but identical information can often be more easily obtained by the use of the *inverse Nyquist diagram*.

Gain and Phase Margins

The *gain margin* is the amount by which the open-loop gain must be increased, at the frequency at which the phase shift is 180°, in order to enclose the -1 point.

The *phase margin* is the additional open-loop phase shift required, at the frequency where the gain is unity, in order that the -1 point shall be enclosed.

Both these quantities may be obtained from the Nyquist diagram, Fig. 6.1. If $G(j\omega)$ is the Nyquist plot, then

$$\text{Gain margin} = \frac{1}{|G(j\omega_0)|}$$

$$\text{Phase margin} = 180° - \underline{/G(j\omega_1)}$$

M and α Circles

A more exact criterion of system performance would be the maximum value of the closed-loop frequency response function $(C/R)(j\omega)$.

For a unity feedback system this is related to the open-loop frequency response function $G(j\omega)$ by the equation

$$\frac{C}{R}(j\omega) = \frac{G(j\omega)}{1 + G(j\omega)}$$

In Cartesian form, $G(j\omega) = x + jy$, and

$$\frac{C}{R}(j\omega) = \frac{x + jy}{1 + x + jy} = M ej^{\alpha}$$

The above equation yields the relationships

$$\left(x + \frac{M^2}{M^2 - 1}\right)^2 + y^2 = \left(\frac{M}{M^2 - 1}\right)^2 \qquad (6.1)$$

$$\left(x + \frac{1}{2}\right)^2 + \left(y - \frac{1}{2\alpha}\right)^2 = \frac{1}{4}\frac{(\alpha^2 + 1)}{\alpha^2} \qquad (6.5)$$

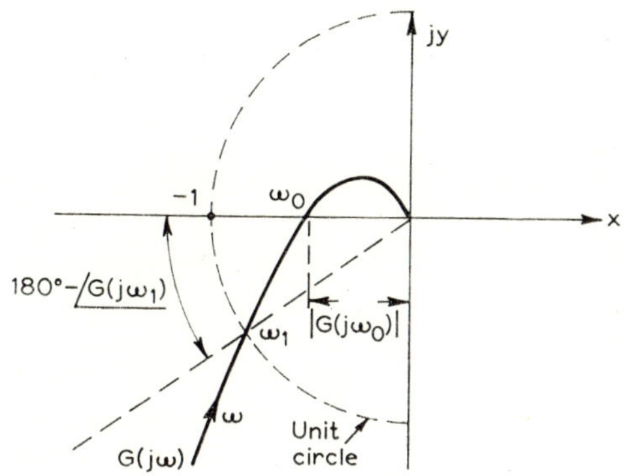

FIG 6.1

Eqn. (6.1), for a constant value of M, is the equation of a circle with centre

$$x = \frac{-M^2}{M^2 - 1}, \quad y = 0$$

and radius

$$\left|\frac{M}{M^2 - 1}\right|$$

As $M = |(C/R)(j\omega)|$ these are a family of circles of constant closed-loop magnitude.

Similarly, eqn. (6.2) describes a family of circles of constant closed-loop phase shift. Typical contours are shown in Fig. 6.2.

Inverse Nyquist Diagram (*Inverse polar plot*)
This is a polar plot of the inverse open-loop transfer function $1/G(j\omega)H(j\omega)$, which possesses certain advantages over the direct plot. These include easier construction of constant M and α loci,

and determination of the effect of the addition of extra feedback paths.

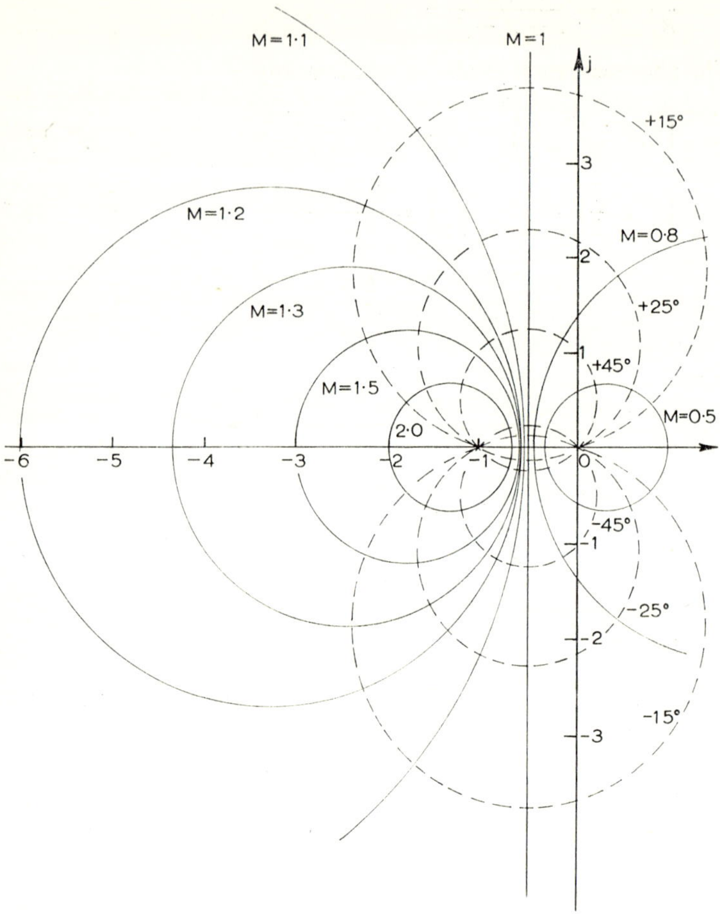

FIG 6.2

—— loci of constant M
- - - loci of constant α

The method of determining the M and α contours is similar to that used in the direct polar plot. Thus

$$\frac{R}{C}(j\omega) = \frac{1}{G(j\omega)} + H(j\omega) \tag{6.3}$$

$$= \frac{1}{M}\,e^{-j\alpha}$$

For the unity feedback condition, $H(j\omega) = 1$, this equation yields the following features:

1. Contours of constant values of M are circles of radius $1/M$ and centre at the -1 point.
2. Contours of constant values of $-\alpha$ are straight lines passing through the -1 point.

Typical contours are shown in Fig. 6.3.

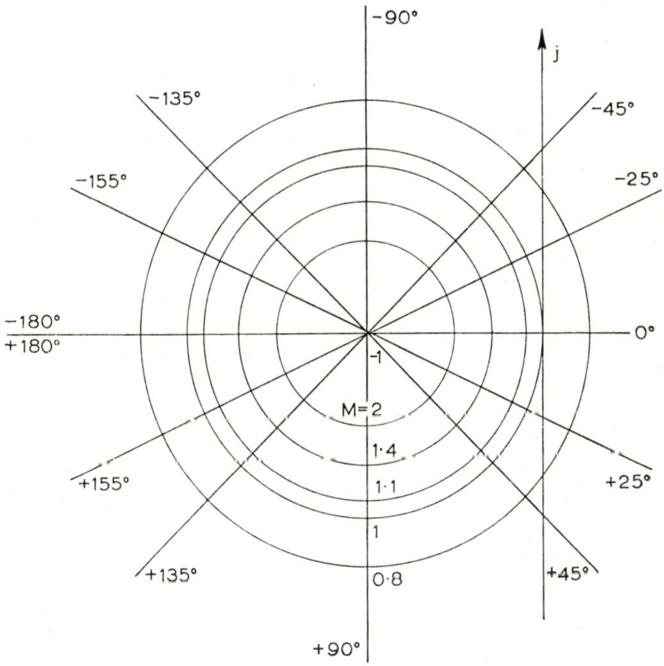

FIG 6.3

Circles are loci of constant M
Radial lines are loci of constant α

Eqn. (6.3) can be used to investigate systems involving multiple-feedback paths. Consider the system shown in Fig. 6.4. Using eqn. (6.3) successively starting from the inner loop, we derive the following relationship:

$$\frac{E}{C}(s) = \left\{\left[\frac{1}{K_1 G_1(s)} + H_1(s)\right] \frac{1}{K_2 G_2(s)} + H_2(s)\right\} \frac{1}{K_3 G_3(s)}$$

which can be written

$$\frac{E}{C}(s) = \frac{1}{K_1K_2K_3}\left\{\frac{1}{G_1(s)\,G_2(s)\,G_3(s)} + \frac{K_1\,H_1(s)}{G_2(s)\,G_3(s)} + \frac{K_1K_2\,H_2(s)}{G_3(s)}\right\} \tag{6.4}$$

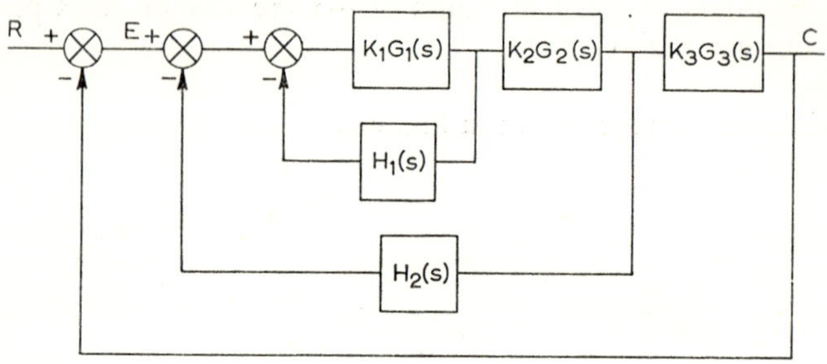

FIG 6.4

As shown later it is easier to plot only the frequency-dependent portion of this equation and to scale the diagram accordingly to take account of the factor $1/K_1K_2K_3$.

Eqn. (6.4) can be remembered by noting that:

1. $K_1K_2K_3$ is the product of the gain terms in outer loop.
2. The first term in the brackets is the reciprocal of the frequency-dependent terms in the outer loop.
3. Successive terms in the brackets are simply the product of the inner loop gain, the feedback path frequency-dependent terms, and the inverse of any succeeding frequency-dependent terms in the outer loop.
4. The number of terms in the brackets is equal to the number of loops.

For an example in the use of this formula see Example 6.7.

In order to illustrate the use of Nyquist and inverse Nyquist plots we consider the voltage regulator problem that has been analysed in earlier chapters.

For an open-loop gain of 50 the block diagram of the system shown in Fig. 5.3 has $K = 0{\cdot}13$. Strictly, it is the signal at the output of the feedback path that is the controlled variable. However, as there is a constant relation between this variable and the generator voltage, any percentage measurements will be identical in both cases. If the absolute value of the generator voltage is required this is $C/0{\cdot}13$.

Suppose we are required to find the maximum value of $|(C/R)(j\omega)|$ and the frequency at which it occurs. We will denote this value by

M_m and the angular frequency at which it occurs as ω_m. The open-loop transfer function is

$$\frac{C}{E}(j\omega) = \frac{50}{(1 + j0{\cdot}05\omega)(1 + j2{\cdot}5\omega)}$$

This can be evaluated directly or with the help of the complex number table given in Appendix 1. Both $(C/E)(j\omega)$ and $(E/C)/(j\omega)$ have been evaluated so that both the Nyquist and the inverse diagrams can be plotted.

ω (rad/s)	0·4	1	2	4	8	16	32
$(1 + j0{\cdot}05\omega)$ $(1 + j2{\cdot}5\omega)$	$1\angle 0°$ $1{\cdot}41\angle 45°$	$1\angle 0°$ $2{\cdot}7\angle 68°$	$1\angle 5°$ $5{\cdot}1\angle 78°$	$1{\cdot}02\angle 11°$ $10\angle 84°$	$1{\cdot}07\angle 21°$ $20\angle 87°$	$1{\cdot}28\angle 38°$ $40\angle 88°$	$1{\cdot}87\angle 58°$ $80\angle 89°$
$(1 + j0{\cdot}05\omega) \times (1 + j2{\cdot}5\omega)$	$1{\cdot}41\angle 45°$	$2{\cdot}7\angle 68°$	$5{\cdot}1\angle 83°$	$10{\cdot}2\angle 95°$	$21{\cdot}4\angle 108°$	$51{\cdot}2\angle 126°$	$149\angle 147°$
$[(1 + j0{\cdot}05\omega) \times (1 + j2{\cdot}5\omega)]^{-1}$	$0{\cdot}71\angle -45°$	$0{\cdot}37\angle -68°$	$0{\cdot}20\angle -83°$	$0{\cdot}10\angle -95°$	$0{\cdot}05\angle -108°$	$0{\cdot}02\angle -126°$	$0{\cdot}006\angle -147°$

These results show only the frequency-dependent portion of the transfer function. In general it is easier to plot only this part of the

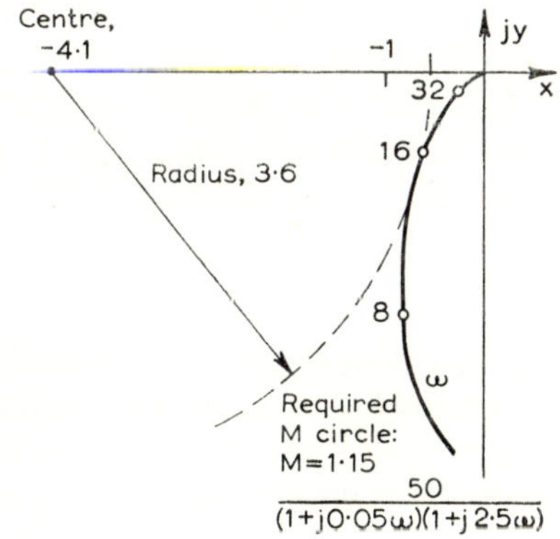

FIG 6.5

function; the constant can be accounted for by a change of scale on the diagram.

The Nyquist diagram is shown in Fig. 6.5, and the inverse Nyquist diagram in Fig. 6.6. In the latter case only the frequency-dependent portion of $(E/C)/(j\omega)$ has been plotted and the diagram has been scaled accordingly.

In order to find M_m from the Nyquist plot it is necessary to find an M circle that is tangential to the plot. The $M = 1{\cdot}15$ circle fulfils this condition.

It is easier to find the required M circle (tangential to the locus) from the inverse Nyquist plot as all the M circles have the same centre. This is shown in Fig. 6.6. The maximum value of M occurs at $\omega = 14$ rad/s; the corresponding value of closed-loop phase shift, α, is approximately 54°.

Alternatively, we may be required to determine the open-loop gain constant to give a desired value of M_m. The loci depicted in Figs. 6.5

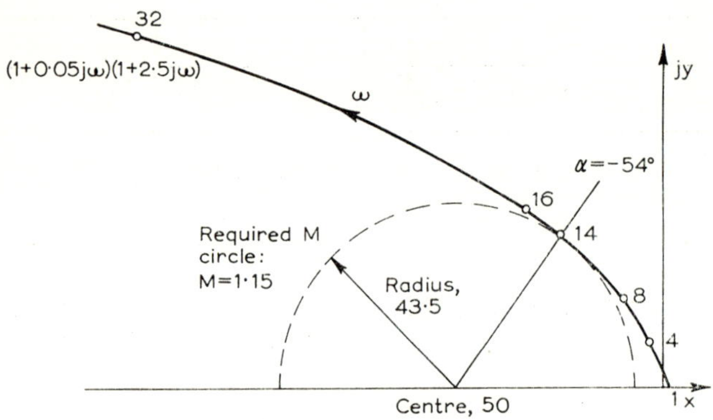

FIG 6.6

and 6.6 could still be obtained but as the open-loop gain is unknown it is not possible to draw the M circles, the scale factor of the diagram being undetermined.

However, the following properties of M circles prove useful for this condition:

Nyquist diagram. Fig. 6.7 (*a*) depicts a typical M circle, having centre on the real axis at $-M^2/(M^2 - 1)$ and radius $M/(M^2 - 1)$. From the right-angled triangle OAB it can be shown that the angle subtended by a tangent, drawn from the origin to the circle at A, and the real axis is given by

$$\psi = \sin^{-1} \frac{1}{M}$$

A perpendicular to the real axis drawn from A meets the real axis at -1. These properties are used in Example 6.5.

Inverse Nyquist diagram. In this case the typical M circle (Fig. 6.7 (*b*)) has radius $1/M$ and centre $(-1, j0)$. A line from the origin tangential to the circle again subtends an angle $\psi = \sin^{-1}(1/M)$ with the real axis. However, the -1 point is the centre of the circle in this case, i.e. a line perpendicular to the tangent at A passes through the point $(-1, j0)$.

The values of M_m and ω_m can be related to those of ω_d and ζ in the time domain.

It is proved in Example 6.2 that the following relationships hold for a second-order system:

$$M_m = \frac{1}{2\zeta\sqrt{(1-\zeta^2)}} \tag{6.7}$$

$$\omega_m = \omega_n\sqrt{(1-2\zeta^2)} \tag{6.8}$$

where ω_n is the natural frequency of the system.

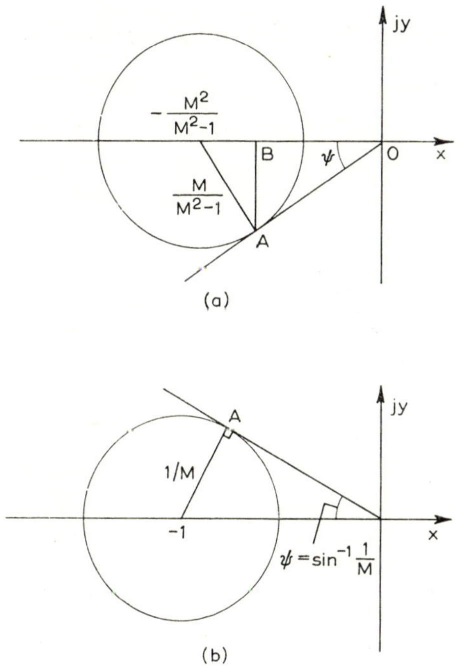

FIG 6.7

Applying these formulae to the voltage regulator example gives $\zeta = 0{\cdot}50$ and $\omega_n = 19{\cdot}9$ rad/s. These compare favourably with the values obtained directly from the characteristic equation of the system in Chapter 3, namely $\zeta = 0{\cdot}49$ and $\omega_n = 20{\cdot}2$ rad/s.

Example 6.1

A servomechanism has the following open-loop frequency response:

ω (rad/s)	2	3	4	5	6	8	10
$\lvert KG(j\omega)\rvert$	7·50	4·80	3·15	2·25	1·70	1·00	0·64
arg $KG(j\omega)$	−118°	−130°	−140°	−149°	−157°	−170°	−180°

If the maximum value of the magnification is to be 1·4 with the loop closed, determine the factor by which the gain must be changed.

Find the gain margin and phase margin for the system when thus altered. (Use the chart provided.) (*Scottish Assoc.*)

Solution

These figures are plotted in Fig. 6.8. For the maximum value of

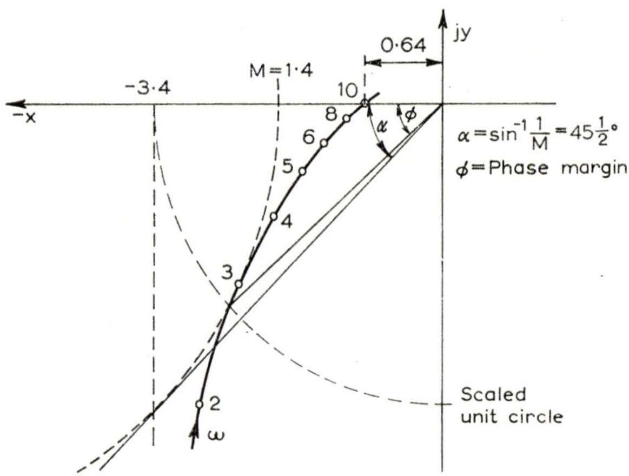

FIG 6.8

magnification, M, to be 1·4 the response plotted must be tangential to the $M = 1{\cdot}4$ circle.

A line is drawn from the origin at angle $\sin^{-1}(1/M)$ and a circle is found with centre on the real axis and tangential to both the plot and this line.

A perpendicular to the real axis from the point of contact of the $M = 1{\cdot}4$ circle and the $\sin^{-1}(1/M)$ line crosses the real axis at −3·4. Hence the gain must be reduced by a factor of 3·4 in order to achieve an M_m of 1·4.

The radius of the unit circle becomes 3·4 in order to measure the phase margin. From Fig. 6.8 the phase margin ϕ is $41\frac{1}{2}°$, and the gain margin is 3·4/0·64, or 5·3.

Example 6.2

Find the range of values of damping ratio ζ for which the system represented by Fig. 6.9 has a peak value of closed-loop frequency

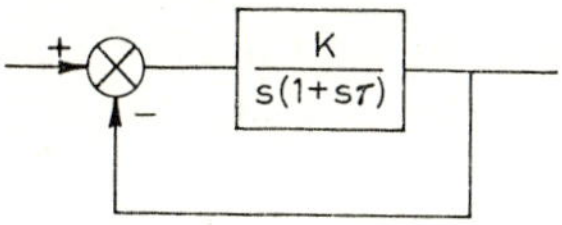

FIG 6.9

response amplitude ratio, $|G(j\omega)|$, greater than unity.

Show that the peak value of $|G(j\omega)|$ is given by

$$|G(j\omega)| = \frac{1}{2\zeta\sqrt{(1-\zeta^2)}}$$

(*Edinburgh Univ.*)

Solution

The closed-loop response is given by

$$\frac{C}{R}(s) = \frac{K}{s(1+s\tau)+K} = \frac{K/\tau}{s^2 + s/\tau + K/\tau}$$

This has the standard form of eqn. (3.5):

$$\frac{C}{R}(s) = \frac{\omega_n^{\ 2}}{s^2 + 2\zeta\omega_n s + \omega_n^{\ 2}}$$

where $\omega_n^{\ 2} = K/\tau$ and $\zeta = \frac{1}{2}/\sqrt{(K\tau)}$.

The closed-loop frequency response is

$$\frac{C}{R}(j\omega) = \frac{\omega_n^{\ 2}}{\omega_n^{\ 2} - \omega^2 + j2\zeta\omega_n\omega}$$

of magnitude

$$\left|\frac{C}{R}(j\omega)\right| = \frac{\omega_n^{\ 2}}{\sqrt{[(\omega_n^{\ 2}-\omega^2)^2 + 4\zeta^2\omega_n^{\ 2}\omega^2]}} \qquad (1)$$

o find the range of values of ζ where $|(C/R)(j\omega)| > 1$ the fol-
ing inequality must hold:

$$\frac{\omega_n^2}{\sqrt{[(\omega_n^2 - \omega^2)^2 + 4\zeta^2\omega_n^2\omega^2]}} > 1$$

$$0 > \omega^2(4\zeta^2\omega_n^2 - 2\omega_n^2 + \omega^2)$$

As ω^2 is always positive,

$$0 > 2(2\zeta^2 - 1) + \left(\frac{\omega}{\omega_n}\right)^2$$

Fulfilment of this condition will depend on the values of ζ and ω. However $(\omega/\omega_n)^2$ is always positive and the condition can never be fulfilled unless

$$(2\zeta^2 - 1) < 0$$

i.e. ζ can take values $0 \to \sqrt{0{\cdot}5}$, i.e. $0 \to 0{\cdot}7$. Under these conditions there will always be a range of values of ω where $|(C/R)(j\omega)| > 1$.

The peak value of $|(C/R)(j\omega)|$ is obtained from the equation

$$\frac{d}{d\omega}\left\{\frac{\omega_n^2}{\sqrt{[(\omega_n^2 - \omega^2)^2 + 4\zeta^2\omega_n^2\omega^2]}}\right\} = 0$$

which yields

$$\omega = \omega_n\sqrt{(1 - 2\zeta^2)}$$

Substituting this value of ω in eqn. (1) gives

$$\left|\frac{C}{R}(j\omega)\right|_{max} = \frac{\omega_n^2}{\sqrt{[(\omega_n^2 - \omega_n^2 + 2\zeta^2\omega_n^2)^2 + 4\zeta^2\omega_n^2\omega_n^2(1 - 2\zeta^2)]}}$$

$$= \frac{1}{2\zeta\sqrt{(1 - \zeta^2)}}$$

Example 6.3

Explain the term "overall or closed-loop transfer function" as applied to a servomechanism.

A servomechanism has the following open-loop frequency response $KG(j\omega)$

ω (rad/s)	1	2	3	4	5	6	8
$\lvert KG(j\omega)\rvert$	4·75	2·70	1·70	0·97	0·63	0·40	0·17
arg $KG(j\omega)$	−104°	−115°	−126°	−138°	−150°	−163°	−197°

Plot the open-loop frequency response.

Find, for the closed-loop system (i) the damped natural frequency for a step input, (ii) the bandwidth of the servomechanism, (iii) the maximum value of M for a sinusoidal input. (*Scottish Assoc.*)

Solution

The values of gain and phase angle are plotted as a Nyquist diagram in Fig. 6.10.

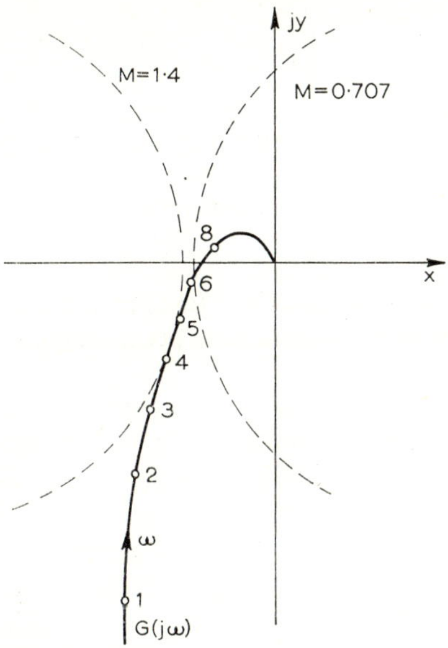

FIG 6.10

By drawing a series of M circles it is found that the $M = 1{\cdot}4$ circle is tangential to the plot at $\omega_m = 4\,\text{rad/s}$. This is the maximum value of M obtainable.

Use of eqns. (6.6) and (6.7) gives $\zeta = 0{\cdot}39$ and $\omega_n = 4{\cdot}8\,\text{rad/s}$. Eqn. (3.7) now gives the damped frequency of oscillation:

$$\omega_d = \omega_n\sqrt{(1 - \zeta^2)} = 4{\cdot}4\,\text{rad/s}$$

The bandwidth of the servomechanism is defined as the frequency at which the magnitude of the closed-loop frequency response is $1/\sqrt{2}$ of its magnitude at $\omega = 0$. As the system appears to be type 1 and there is unity feedback its closed-loop gain at $\omega = 0$ can be taken as 1. A gain of $1/\sqrt{2}$ times the gain at $\omega = 0$ gives an M circle of $M = 0{\cdot}707$. This is drawn on Fig. 6.10 and intersects the plot at $\omega = 6\,\text{rad/s}$.

Example 6.4

State the rigorous form of Nyquist's theorem.

A process with proportional control can be approximated by the following open-loop transfer function:

$$\frac{0{\cdot}2\,e^{-2s}}{s(1 + s)(1 + 4s)}$$

time-constants being in minutes.

Plot the complete Nyquist locus and determine the gain and phase margins. Would the control be stable? If stable, would you regard the margin of stability as acceptable? (*Nottingham Univ.*)

Solution

The magnitude and angle of the open-loop transfer function are evaluated in the following table (ω is in radians per minute).

$s = j\omega$	0	0·1	0·2	0·3	0·5
e^{-2s}	$1\angle 0°$	$1\angle -11\frac{1}{2}°$	$1\angle -23°$	$1\angle -34\frac{1}{2}°$	$1\angle -57\frac{1}{2}°$
s	$0\angle 90°$	$0{\cdot}1\angle 90°$	$0{\cdot}2\angle 90°$	$0{\cdot}3\angle 90°$	$0{\cdot}5\angle 90°$
$1 + s$	$1\angle 0°$	$1{\cdot}0\angle 5\frac{1}{2}°$	$1\angle 11°$	$1\angle 13\frac{1}{2}°$	$1{\cdot}1\angle 26°$
$1 + 4s$	$1\angle 0°$	$1{\cdot}1\angle 21\frac{1}{2}°$	$1{\cdot}3\angle 38°$	$1{\cdot}6\angle 50°$	$2{\cdot}2\angle 63\frac{1}{2}°$
$s(1 + s)(1 + 4s)$	$0\angle 90°$	$0{\cdot}11\angle 117°$	$0{\cdot}26\angle 139°$	$0{\cdot}48\angle 153°$	$1{\cdot}21\angle 180°$
$\dfrac{0{\cdot}2e^{-2s}}{s(1 + s)(1 + 4s)}$	$\infty\angle -90°$	$1{\cdot}82\angle 128\frac{1}{2}°$	$0{\cdot}77\angle -162°$	$0{\cdot}42\angle -188°$	$0{\cdot}16\angle -237°$

Fig. 6.11 is a sketch of the complete Nyquist plot, and Fig. 6.12 is a more accurate plot of the region below $\omega = 0{\cdot}5$ rad/min. From Fig. 6.12 the phase margin $\phi = 25°$ and the gain margin $1/a = 1/0{\cdot}5 = 2$. The system is stable but the phase margin is not acceptable.

Example 6.5

Explain what is meant by M circles in relation to the direct or inverse Nyquist diagram.

A control system has the following open-loop frequency response where ω is the angular frequency of the test signal in radians per second, $|G|$ is the modulus of the ratio of the output to the input and ϕ is the phase angle between input and output in degrees.

ω	1·0	1·5	2·0	3·0	5·0
$\|G(j\omega)\|$	3·58	2·13	1·40	0·72	0·12
ϕ	−123°	−135°	−146°	−163°	−190°

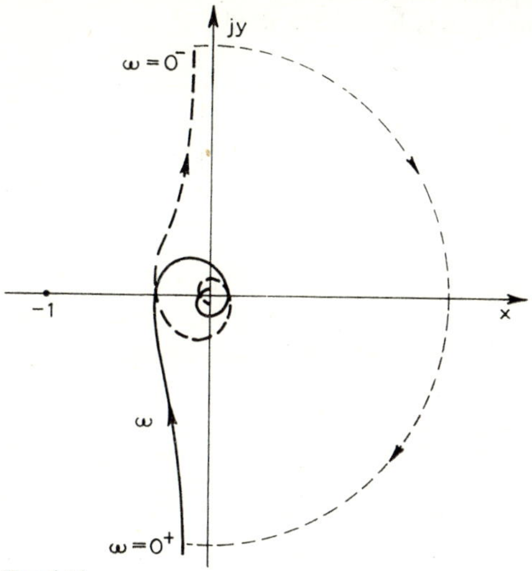

FIG 6.11

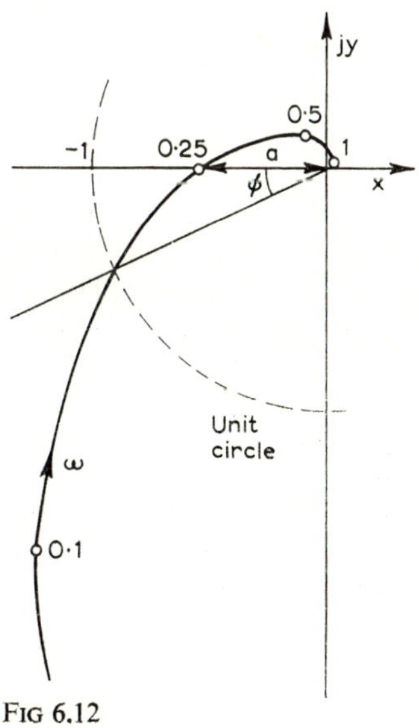

FIG 6.12

Write down an expression showing the probable form of the open-loop transfer function.

Determine the value of the loop gain constant for the value of M_{max} to be between 1·3 and 1·4. (*Cardiff*)

Solution

The inverse Nyquist plot is shown in Fig. 6.13. In order to plot this diagram the following figures have been used:

ω (rad/s)	1·0	1·5	2·0	3·0	5·0
$1/\lvert G\rvert$	0·28	0·47	0·71	1·39	8·3
$-\phi$	123°	135°	146°	163°	190°

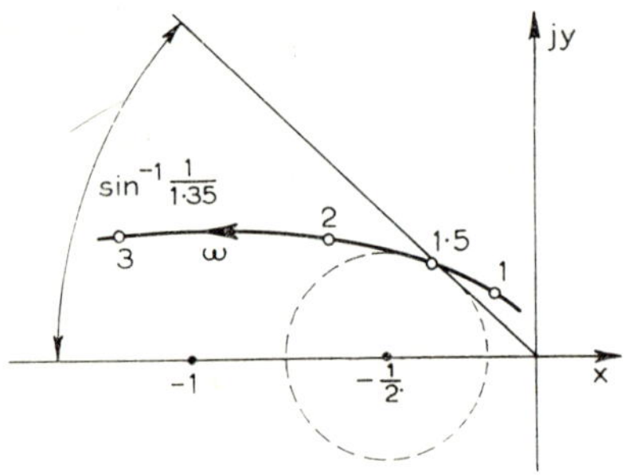

FIG 6.13

In order to locate an M circle for M_{max} to be 1·35, a line is drawn, as shown at an angle $\sin^{-1}(1/1{\cdot}35)$ (48°) to the real axis. An M circle is now constructed, whose centre is on the real axis and which is tangential to both the plot and the "$\sin^{-1}(1/M)$" line. The centre of this circle is at $-\frac{1}{2}$ so the plot must be multiplied by a factor of 2, i.e. $K = 0{\cdot}5$ for the centre to become the -1 point. (This can be checked by plotting $1/0{\cdot}5G$ and noting that the $M = 1{\cdot}35$ circle is tangential to this plot.)

As the plot appears to approach the imaginary axis tangentially this suggests an integrator in the transfer function. At $\omega = 5{\cdot}0$ rad/s the phase shift is 190°; hence there must be at least two exponential lags in the system. However, as the phase shift does not increase very rapidly, there are not likely to be more than two.

The probable form of the transfer function is

$$G(s) = \frac{K}{s(1 + s\tau_1)(1 + s\tau_2)}$$

Example 6.6

The open-loop frequency response of a control system is given in the table.

ω (rad/s)	1·0	1·5	2·0	2·5	3·0	3·5	4·0
$\lvert G\rvert$	6·7	2·8	1·8	1·2	0·9	0·7	0·5
ϕ	−130°	−140°	−148°	−153°	−158°	−160°	−164°

Plot an inverse Nyquist diagram and determine the maximum value of the closed-loop response M and the frequency at which it occurs.

If stabilizing feedback is applied in the form of a rate of change of output and its numerical value is $j0{\cdot}15\omega$, determine graphically the resultant open-loop response locus, justifying the construction and the new value of M. (*Cardiff*)

Solution

The figures for $1/\lvert G\rvert$ and $-\phi$ are given in the following table:

ω (rad/s)	1·0	1·5	2·0	2·5	3·0	3·5	4·0
$1/\lvert G\rvert$	0·15	0·36	0·55	0·83	1·1	1·43	2·0
$-\phi$	130°	140°	148°	153°	158°	160°	164°

These figures have been plotted as an inverse Nyquist diagram in Fig. 6.14. Construction of the appropriate M circle gives M_m as $1/0{\cdot}4 = 2{\cdot}5$ at an angular frequency of 2·8 rad/s.

With the stabilizing feedback present we have

$$\frac{E}{C}(j\omega) = \frac{1}{G(j\omega)} + H(j\omega)$$

$H(j\omega)$ is $0{\cdot}15j\omega$ and is tabulated as follows:

ω	1·0	1·5	2·0	2·5	3·0	3·5	4·0
$H(j\omega)$	$0{\cdot}15\underline{/90^\circ}$	$0{\cdot}23\underline{/90^\circ}$	$0{\cdot}3\underline{/90^\circ}$	$0{\cdot}38\underline{/90^\circ}$	$0{\cdot}45\underline{/90^\circ}$	$0{\cdot}53\underline{/90^\circ}$	$0{\cdot}6\underline{/90^\circ}$

These vectors can be added to the original plot to give the new plot of $(E/C)(j\omega)$. Construction of the appropriate M circle gives

$$M = \frac{1}{0{\cdot}78} = 1{\cdot}28 \text{ at } \omega = 2{\cdot}3\,\text{rad/s}$$

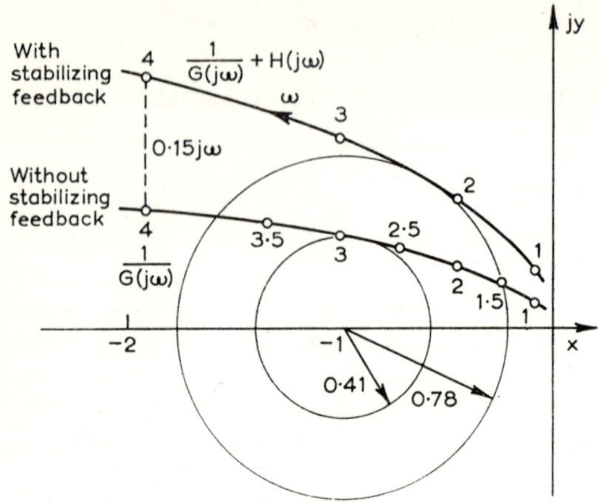

FIG 6.14

Example 6.7

A precise fast-acting position control system is required as an analogue computer servo-multiplier. The scheme to be used is shown in Fig. 6.15.

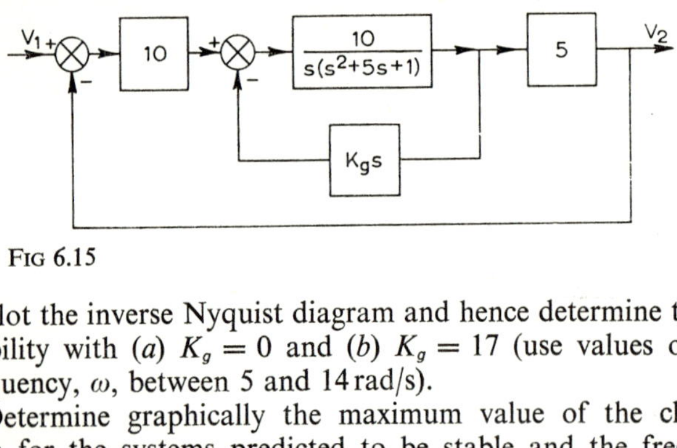

FIG 6.15

Plot the inverse Nyquist diagram and hence determine the system stability with (*a*) $K_g = 0$ and (*b*) $K_g = 17$ (use values of angular frequency, ω, between 5 and 14 rad/s).

Determine graphically the maximum value of the closed-loop gain for the systems predicted to be stable and the frequency at which it occurs. (*Southampton Univ.*)

Solution

The block diagram can be redrawn as shown in Fig. 6.16. Using eqn. (6.4),

$$\frac{E}{V_2} = \frac{1}{500}\left\{s(s^2 + 5s + 1) + 100\frac{K_g}{10}s\right\}$$

When $K_g = 0$, E/V_2 can be plotted from the following table:

ω (rad/s)	5	8	10	12	14
$s = j\omega$	$5\angle 90^\circ$	$8\angle 90^\circ$	$10\angle 90^\circ$	$12\angle 90^\circ$	$14\angle 90^\circ$
$s^2 + 5s + 1$ $=(1 - \omega)^2 + 5j\omega$	$35{\cdot}8\angle 133^\circ$	$75{\cdot}5\angle 148^\circ$	$112\angle 153^\circ$	$154\angle 157^\circ$	$207\angle 161^\circ$
$s(s^2 + 5s + 1)$	$174\angle 223^\circ$	$604\angle 238^\circ$	$1120\angle 243^\circ$	$1848\angle 247^\circ$	$2900\angle 251^\circ$

The quadratic factor $1 + 5s + s^2$ can be evaluated by writing the expression in the form

$$(1 - \omega^2)\left(1 + \frac{5}{1 - \omega^2}j\right)$$

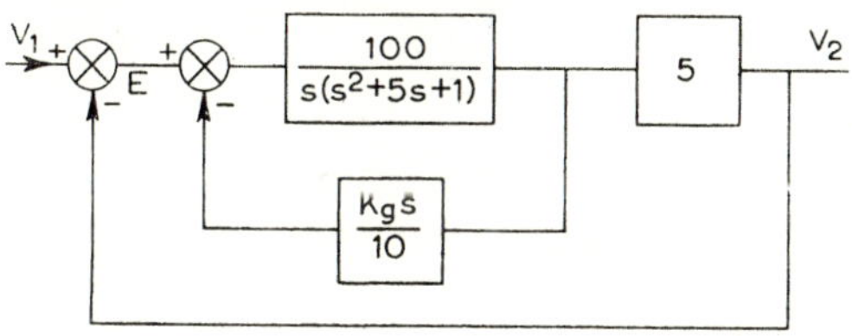

FIG 6.16

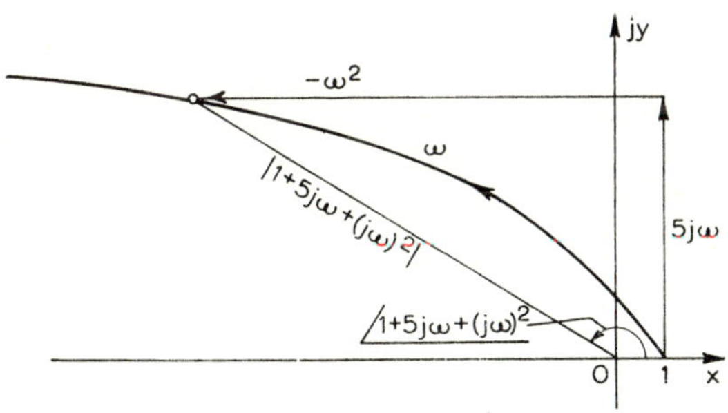

FIG 6.17

and making use of the complex number table. Alternatively the factor can be determined graphically as in Fig. 6.17.

The plot is shown in Fig. 6.18: the system is unstable. When $K_g = 17$ the relationship between E and V_2 is

$$\frac{E}{V_2} = \frac{1}{500}\{s(s^2 + 5s + 1) + 170s\}$$

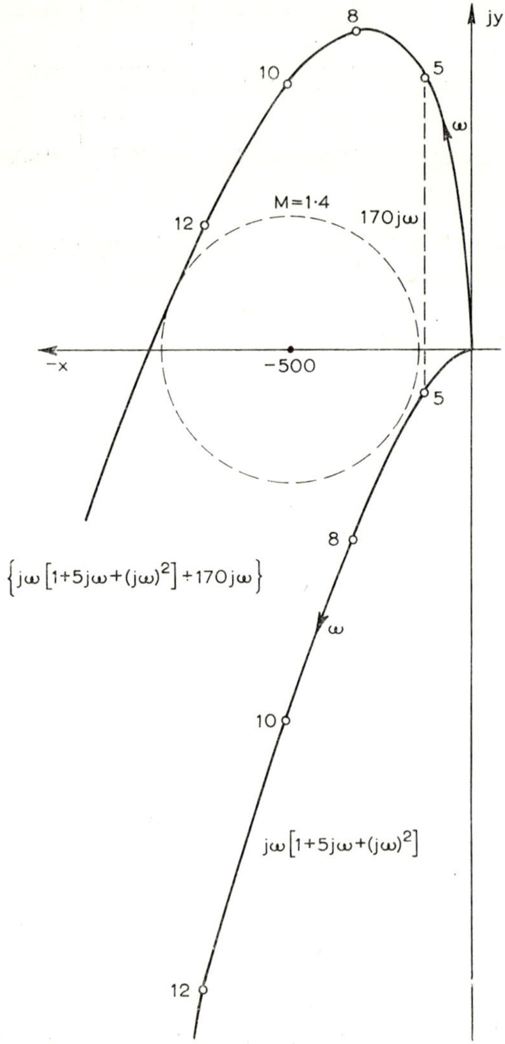

FIG 6.18

hence the factor $170j\omega \equiv 170\omega\underline{/90^\circ}$ must be added vectorially to the original plot. The system is now stable with a maximum value of M equal to 1·4 at an angular frequency of 13 rad/s.

For the system to become unstable the crossing point of the real axis would be the -1 point in the unscaled plot. This gives a gain of $\approx$870.

Example 6.8

A feedback system having the forward-path transfer function

$$\frac{20}{(1 + 0{\cdot}5s)(1 + 2s)}$$

is considered to have unity transfer function in the monitoring feedback path. Obtain the maximum value of M, the magnitude of the closed-loop frequency response, for the system and the angular frequency at which it occurs.

It is later found that the monitoring feedback elements introduced an exponential lag of time-constant 50 ms. What effect does this have on the performance of the system in terms of the above quantities?

Note. Do not extend the plot beyond an angular frequency of 6 rad/s. (*Lanchester*)

Solution

The inverse of the forward path transfer function is

$$0{\cdot}05(1 + 0{\cdot}5s)(1 + 2s)$$

and the frequency-dependent portion of this expression is as follows:

ω (rad/s)	0	1·0	2·0	3·0	4·0	6·0
$1 + 0{\cdot}5j\omega$	$1\underline{/0}$	$1{\cdot}12\underline{/26\frac{1}{2}^\circ}$	$1{\cdot}41\underline{/45^\circ}$	$1{\cdot}8\underline{/56\frac{1}{2}^\circ}$	$2{\cdot}2\underline{/63\frac{1}{2}^\circ}$	$3{\cdot}2\underline{/71\frac{1}{2}^\circ}$
$1 + 2j\omega$	$1\underline{/0}$	$2{\cdot}23\underline{/63\frac{1}{2}^\circ}$	$4{\cdot}1\underline{/76^\circ}$	$6{\cdot}1\underline{/80\frac{1}{2}^\circ}$	$8\underline{/83^\circ}$	$12\underline{/85^\circ}$
$(1 + 0{\cdot}5j\omega)(1 + 2j\omega)$	$1\underline{/0}$	$2{\cdot}5\underline{/90^\circ}$	$5{\cdot}8\underline{/121^\circ}$	$11{\cdot}0\underline{/137^\circ}$	$17{\cdot}6\underline{/146\frac{1}{2}^\circ}$	$384\underline{/156\frac{1}{2}^\circ}$

The open-loop inverse polar plot is shown in Fig. 6.19, and the value of M_m is found to be 1·8 at $\omega = 4$ rad/s. To save labour, the diagram has been scaled by a factor of 20 in the usual manner. The effect of the non-unity feedback path is determined by the use of eqn. (6.3):

$$\frac{R}{C}(j\omega) = \frac{1}{G(j\omega)} + H(j\omega)$$

Hence to the plot of $1/G(j\omega)$ must be added a plot of $H(j\omega)$. However, it must be noted that the resultant locus is now the inverse of the closed-loop response.

$$\frac{R}{C}(j\omega) = \frac{(1 + 0{\cdot}5j\omega)(1 + 2j\omega)}{20} + \frac{1}{1 + 0{\cdot}05j\omega}$$

$$= \frac{1}{20}\left\{(1 + 0{\cdot}5j\omega)(1 + 2j\omega) + \frac{20}{1 + 0{\cdot}05j\omega}\right\}$$

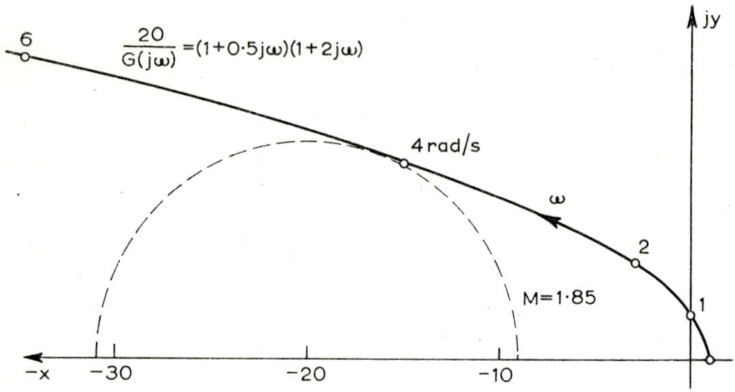

FIG 6.19

The values of the additional term in this expression are tabulated for the range of ω as shown.

ω (rad/s)	1·0	2·0	3·0	4·0	6·0
$1 + 0{\cdot}05j\omega$	$1\underline{/0}$	$1\underline{/5\frac{1}{2}^\circ}$	$1\underline{/8\frac{1}{2}^\circ}$	$1{\cdot}02\underline{/11^\circ}$	$1{\cdot}04\underline{/16\frac{1}{2}^\circ}$
$\dfrac{20}{1 + 0{\cdot}05j\omega}$	$20\underline{/0}$	$20\underline{/-5\frac{1}{2}^\circ}$	$20\underline{/-8\frac{1}{2}^\circ}$	$19{\cdot}6\underline{/-11^\circ}$	$19{\cdot}2\underline{/-16\frac{1}{2}^\circ}$

These figures are added vectorially to the original plot to give that shown in Fig. 6.20.

The maximum value of M is determined by finding the smallest circle centred on the *origin*, which is tangential to this plot. The maximum value of M is 3 and occurs at an angular frequency of approximately 4·5 rad/s. Hence the effect of introducing the exponential lag into the feedback path is to cause a considerable increase in the peak value of the closed-loop frequency response. This is indicative of a decrease in the damping of the transient response.

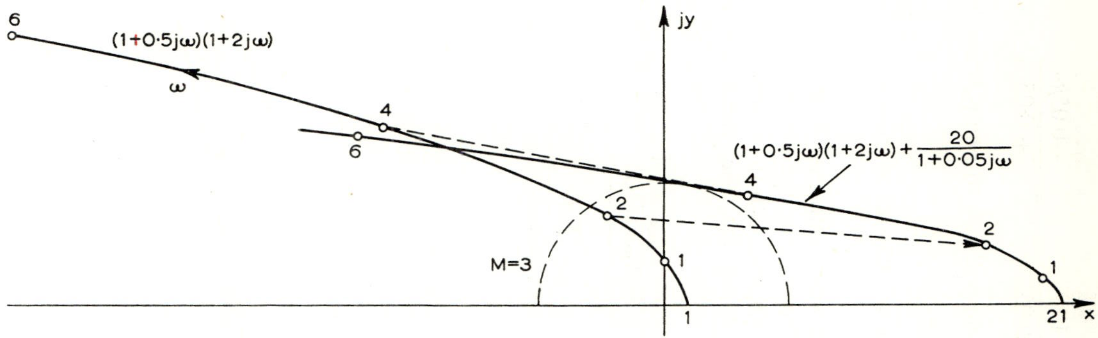

FIG 6.20

An alternative approach to this problem is to split the feedback path into two parts, one of which is unity. Thus $H(j\omega)$ is replaced by $1 + H'(j\omega)$. Here

$$H'(j\omega) = \frac{1}{1 + 0{\cdot}05j\omega} - 1 = \frac{-0{\cdot}05j\omega}{1 + 0{\cdot}05j\omega}$$

The modified system is shown in Fig. 6.21 and is based upon the equation

$$\frac{E}{C}(j\omega) = \frac{1}{20}\left[(1 + 0{\cdot}05j\omega)(1 + 2j\omega) - \frac{j\omega}{1 + 0{\cdot}05j\omega}\right]$$

As the original path was subtracted from R, the subsidiary path becomes positive. The inverse open-loop plot is obtained in the manner of Example 6.6, but with the exception that the vector due to the term $j\omega/(1 + j\omega)$ has the effect of bringing the final

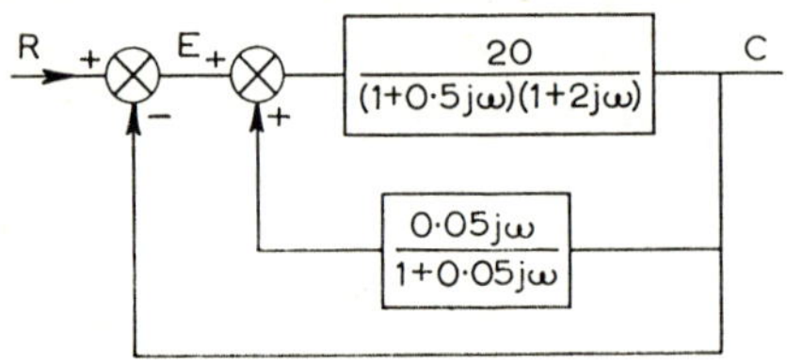

FIG 6.21

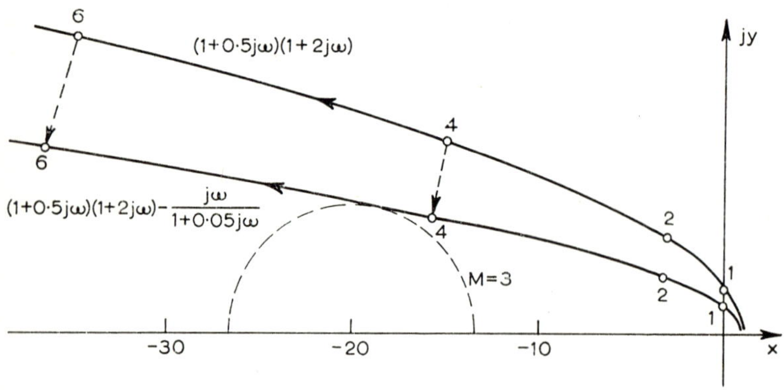

FIG 6.22

locus nearer the -20 point, i.e. decreasing the stability of the system (see Fig. 6.22). The value of M_m is determined by M circles centred on the -20 point.

PROBLEMS

6.1 Discuss the practical application of Nyquist's criterion for the prediction of the closed-loop stability of a feedback control system. Give sketches of typical Nyquist plots and quote examples of practical systems having such plots. State whether the method has any disadvantages.

An industrial process has an open-loop transfer function of the form

$$G(s) = \frac{10}{(1 + 0{\cdot}03s)^3}$$

Sketch on squared paper a polar plot of open-loop frequency response over the range 8 Hz to 12 Hz. What do you deduce from the form of this plot?

A compensating unit having the response shown in the table below is now inserted in the feedback loop. Plot and comment on the new open-loop polar frequency response. Estimate and comment on the resulting system gain and phase margin.

Frequency (Hz)	8	10	12
Phase shift	17°	14°	13°
Amplitude gain	0·84	0·88	0·90

(*London Univ.*)

Ans. Unstable; 1·33, $12\frac{1}{2}$°.

6.2 Explain and justify the Nyquist criterion of stability when applied to a feedback system having only one feedback loop. Sketch the transfer locus for the open-loop transfer function

$$G(s) = \frac{K}{s(\tau_1 s + 1)(\tau_2 s + 1)}$$

and explain the effect of the value of K on the stability of the system when the loop is closed. What effect will the ratio of τ_1 to τ_2 have on system stability, and how will it affect the value of K? (*Cardiff*)

6.3 What are the characteristics of an unstable feedback system, and what is the general relationship determining the condition for stability?

The forward path transfer operator of a system is given by

$$KG(s) = \frac{12}{(5 + s)(1 + s + s^2)}$$

where $\theta_o = KG(s)\,\theta_e$, and $s = d/dt$.

Plot the polar diagram for the system and determine the frequency of oscillation at which the phase lag between θ_o and θ_e is π radians. Use values of $\omega = 0$, 0·8, 1·2, 2·0, 3·0, ∞. (*Nottingham*)

Ans. 2·5 rad/s.

6.4 A control system having unity feedback was found to have an insufficiently damped response and series compensation was incorporated. Frequency response tests were carried out both before and after compensation and some of the results are given:

Before compensation, ω (rad/s)	15	17	19	25
Gain	1·10	0·884	0·722	0·432
Phase angle	−160°	−163°	−166°	−174°
After compensation, ω (rad/s)	22	26	30	33
Gain	1·115	0·883	0·715	0·615
Phase angle	−136°	−143°	−149°	−153°

Plot these results and determine the values of M_m before and after compensation. Sketch roughly the response of the system to a step input before and after compensation, using the same axes for the two curves. (*Brunel Univ.*)

Ans. 3·0, 1·45.

6.5 The following experimental results were obtained from an open-loop frequency-response test of an automatic control system:

Angular frequency (rad/s)	3	4	5	6	8	10	12
Gain	0·95	0·66	0·480	0·362	0·230	0·150	0·103
Phase angle	−123°	−134°	−143°	−152°	−167°	−180°	−190°

(i) Plot the locus of the loop transfer function.
(ii) Measure the gain and phase margins.
(iii) By what factor should the gain be increased so that the maximum closed-loop gain is 1·4, and what would then be the gain and phase margin? (*Brunel Univ.*)

Ans. (ii) 6·7, 59°; (iii) 1·75, 3·8, 42°.

6.6 (*a*) Explain briefly the use of a Nyquist diagram in examining the stability of servo systems.

(*b*) A tachogenerator is geared to the output shaft of a servo without auxiliary feedback. The e.m.f. produced is 5 V when the output shaft speed is 1 rad/s. By plotting a Nyquist diagram of the forward transfer function of the position servo, estimate the sensitivity of the output shaft position transducer taking the value of M_{max} for the closed-loop servo to be 1·3. The following data are obtained from a resolved components indicator connected to the tachogenerator. The phase reference and the input to the opened loop at the error-voltage point is a sinusoidal wave of constant amplitude 1 V r.m.s., whose angular frequency, ω, may be varied.

ω (rad/s)	1·6	2·0	2·8	4·0	8·0
V_{ref} (mV r.m.s.)	+246	+222	+172	+106	0
V_{quad} (mV r.m.s.)	−128	−148	−172	−176	−120

(*I.E.R.E.*)

Ans. (*b*) 60 V/rad.

6.7 The equation governing a control system is $K\theta_e = s(1 + \tau s)\theta_o$, where

θ_e is the error in radians,
θ_o is the output position in radians,
K is the steady state gain of the system,
τ is the system time-constant.

If $\omega_o = 4$ rad/s and the damping ratio is 0·5, determine the values of (*a*) K, (*b*) τ, (*c*) steady-state velocity error for $\omega_i = 20°$/s, (*d*) the maximum value of the magnification, M_{max}. Sketch roughly the Nyquist plot. (*Brunel Univ.*)

Ans. (*a*) $K = 4$, (*b*) $\tau = 0·25$, (*c*) 5°, (*d*) $M_{max} = 1·15$.

6.8 A type-one control system has two simple time lags both having time-constants of one second.

(i) Determine the open-loop transfer function of the system in terms of the velocity error constant K for a steady sinusoidally varying error of angular frequency ω.
(ii) Plot the inverse frequency locus for this transfer function for a value of $K = 1$ rad per rad/s. (Use angular frequencies of 0·3, 0·4, 0·5, 0·7, 1·0 and 1·2 rad/s and a scale of 2 inches to 1 unit.) Determine the gain margin and phase margin from the locus diagram, and comment on the stability of the system when on closed-loop.
(iii) A tachogenerator with an effective constant of 0·7 V/rad-s^{-1} is used to provide parallel compensation for the system. Construct the new inverse open-loop frequency locus of the compensated system and determine the new gain margin and

phase margin. Comment on the closed-loop stability of the compensated system and briefly explain the theory underlying your construction. (*I.E.R.E.*)

Ans. (ii) 6 dB, 21°, system absolutely but not relatively stable; (iii) 10·4 dB, 52°, system absolutely and relatively stable.

6.9 Determine the transfer function of a control element which has a pure time delay of τ_V seconds.

If a control element has such a transfer function followed by a simple exponential delay, sketch the $G(j\omega)$ locus. Determine the frequency at which the phase lag is 180° if the time-constants are equal. What is the attenuation when a sinusoidal signal of this frequency is applied to the element? (*Nottingham*)

Ans. $\omega = 2/\tau$, 0·45.

6.10 (*a*) Explain the meaning of the terms (i) transfer function, (ii) frequency response, as applied to control systems, and briefly relate them to the closed-loop response.

(*b*) For a simple second-order system, compare the effect of the damping ratio ζ on the peak overshoot of the transient response and on the value of M_{max}, and comment on the value of ζ for each to just cease to exist.

(*c*) The centres and radii of some M-circles are given below. If a particular system has the following open-loop frequency response, plot from the results its closed-loop θ_o/θ_i against ω on linear graph paper.

M circles

M	Centre	Radius
1·2	−3·27	2·73
1·3	−2·45	1·88
1·4	−2·04	1·46
1·5	−1·8	1·2
1·6	−1·64	1·03

System

ω	4	8	10	13	15	18	20	21
$\lvert KG\rvert$	2·35	1·8	1·5	1·25	0·95	0·7	0·65	0·4
$\angle KG$	124·5°	131°	134°	138°	143·5°	148·5°	151·5°	180°

(*Brunel Univ.*)

6.11 The closed-loop frequency response of a control system may be expressed as

$$\frac{\theta_o}{\theta_i}(j\omega) = M\underline{/\tan^{-1} N}$$

Show that, for a system having a forward transfer function $KG(j\omega)$ and unity feedback, the loci of $KG(j\omega)$ for fixed values of M are circles.

Experimental data for an inverse Nyquist diagram are given below

Frequency	ω_1	ω_2	ω_3	ω_4	ω_5	ω_6	ω_7	ω_8	ω_9
$\frac{1}{KG(j\omega)}$	0·59 +110°	0·79 +123°	1·03 +142°	1·27 +163°	1·61 180°	2·33 +199°	4·17 +214°	10 231°	∞ 270°

Plot the inverse Nyquist diagram and evaluate the maximum value of the magnification M_m. (*Liverpool Univ.*)

Ans. 2·3.

6.12 Show clearly how the closed-loop gain of a system may be derived graphically from a frequency locus of the open loop transfer function $G(j\omega)$.

A position controller has unity feedback, and a forward path transfer function

$$G(j\omega) = \frac{0{\cdot}8}{j\omega(j\omega + 0{\cdot}6)}$$

Find the maximum magnitude of the overall gain and the frequency at which it occurs. (*Swansea*)

Ans. 1·58, $\omega = 0{\cdot}8$.

6.13 Explain the terms "type O," "type 1," "type 2," etc. The system in Fig. 6.23 is of which type? How can it be modified to make it of a higher type?

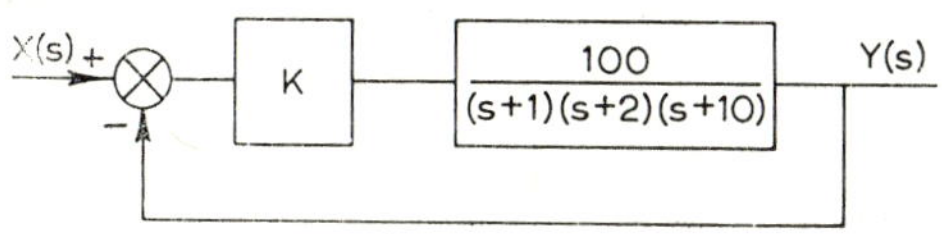

FIG 6.23

If $K = 0{\cdot}5$, what is the steady-state error when a step input $X(s) = 10/s$ is applied?

Draw the Nyquist plot for this system and measure the gain and phase margins. (*Swansea*)

Ans. Type 0, 2·8, 8, 72°.

6.14 (*a*) Explain in detail what is meant by a frequency response test and describe the principles of operation of any pieces of equipment which can be used to make such a test on a d.c. servomechanism.

(*b*) An open-loop frequency response test was performed on a d.c. servomechanism and a Nyquist diagram of the form shown in Fig. 6.24 was obtained. Deduce the open-loop transfer function of

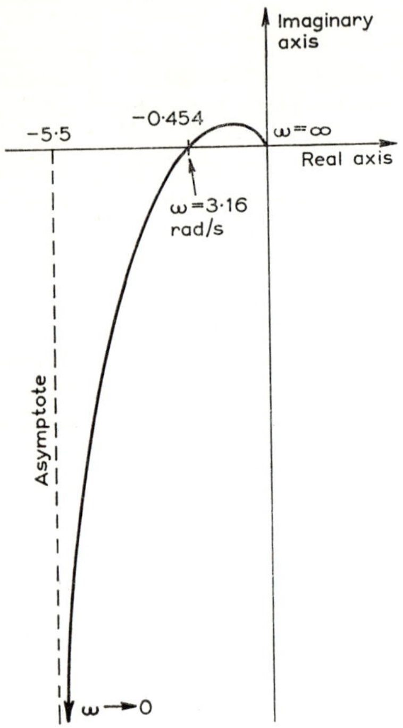

FIG 6.24

the system in terms of the operator $j\omega$, where ω is the angular frequency. (*I.E.R.E.*)

Ans. (*b*) $\dfrac{5}{j\omega(1 + j\omega)(1 + 0{\cdot}1j\omega)}$

6.15 A position control system has the forward-loop transfer function

$$\frac{2}{s(1 + s)(1 + 0{\cdot}1s)}$$

unity feedback being employed. Plot the complete Nyquist contour. Determine the gain and phase margins. (*Nottingham Univ.*)

Ans. 4·5, 34°.

6.16 The transfer function of a process from controlling to controlled variable has been estimated as

$$\frac{e^{-s/3}}{(1+s)^2}$$

for which the unit of time is 1 hour.

Using frequency-response methods, calculate the gain factor of a proportional controller to give an acceptable margin of stability. What form of two-term controller could be employed to achieve a faster response? Estimate the time-constant associated with the setting of such a two-term controller. (*Nottingham Univ.*)

Ans. For $M_m = 1{\cdot}2$, $K = 2{\cdot}6$, $\tau = 1$ s.

6.17 Derive the expression giving the locus, on the Nyquist chart of points of equal amplitude, on the closed loop frequency response curve (M-contours), assuming unity negative feedback.

Describe how to estimate the change of gain required to ensure that any given system with unity feedback shall have a particular maximum amplitude response when subjected to a closed-loop frequency response test. (*Loughborough*)

6.18 A position control system has a forward transfer function

$$KG(s) = \frac{100}{s(s+10)}$$

and a feedback transfer function

$$H(s) = 1 + \tau s$$

For $\tau = 0$ calculate the damping ratio, the undamped natural frequency, and the velocity lag when $\theta_i = 5t$ radians. Determine the magnification M and the phase angle between θ_i and θ_o when θ_i is a sinusoidal signal of frequency $10/2\pi$ hertz for $\tau = 0$.

Calculate the value of τ required for critical damping and the value of τ for which the system will hunt. (*Liverpool Univ.*)

Ans. $\zeta = 0{\cdot}5$, $\omega_n = 10$, 0·5 rad; $M = 1$, $-90°$. For critical damping $\tau = 0{\cdot}1$; for $\zeta = 0{\cdot}7$, $\tau = 0{\cdot}04$.

6.19 Discuss the classification of closed-loop control systems in terms of error conditions. Sketch a typical Nyquist locus for a class

2 control system and explain how the numerical value of the error constant, K_2, may be determined.

Plot the inverse Nyquist loci (using values of ω between 0 and 2) for the system whose block diagram is given in Fig. 6.25, and deter-

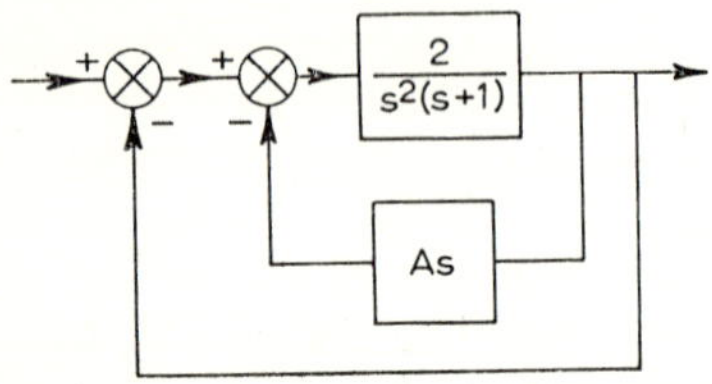

FIG 6.25

mine a value for A which will make M_{max} equal to 1·4. When A has this value, what is the value of the system resonant frequency? (*Rugby*)

Ans. 1·8, 0·29 Hz.

6.20 Discuss the use of the peak magnification, M_{max}, criterion in the design of closed-loop control systems from open-loop frequency-response data.

Show that the curves of constant inverse closed-loop magnitude, M^{-1}, associated with an open-loop inverse polar frequency locus are circles centred at $(-1, j0)$; and find their radius.

Data for the open-loop inverse frequency response of a particular system, whose steady-state error coefficient is unity, are given below:

ω (rad/s) $\{\phi(j\omega)\}^{-1}$	0 0	0·2 $-0·09+j0·19$	0·3 $-0·19+j0·28$	0·4 $-0·36+j0·35$
ω (rad/s) $\{\phi(j\omega)\}^{-1}$	0·5 $-0·55+j0·38$	0·6 $-0·77+j0·39$	0·8 $-1·06+j0·2$	1·0 $-1·16-j0·1$

When the steady-state error coefficient is set to give $M_{max} = 1·3$, find the resulting steady-state error of the system for (*a*) a unit constant input, (*b*) a unit constant-velocity input, (*c*) a unit constant-acceleration input. Find also the gain and phase margins and the resonant frequency. (*Cambridge Univ.*)

Ans. (*a*) 0; (*b*) 2; (*c*) ∞; 2·3, 44°, 0·4 rad/s.

6.21 Obtain an expression so that the stability of the system described in Problem 2.8 may be investigated by the use of the inverse-polar diagram.

$$\textit{Ans.}\quad E/\theta_i = \frac{1}{5K}\{s(1 + 0{\cdot}1s)(1 + s) + 2Ks + 15Ks^2\}$$

6.22 A, B and C are three fixed points along the path of a conveyor belt, the belt passing the points in the order given. Material in powdered form is fed on to the belt at A by a feed hopper, the rate of feed (weight delivered per second) $m(t)$ being controlled by a valve. The material between B and C is weighed continuously, and a signal proportional to the difference between the desired weight W_1 and the actual weight $W(t)$ at any instant controls the hopper valve, so that

$$\frac{d}{dt}m(t) = K\{W_1 - W(t)\}$$

If AB $= b$, BC $= d$, and the linear speed of the belt is V, show that the system just becomes unstable if

$$K = \frac{\omega_c{}^2}{2\sin\left(\dfrac{\omega_c d}{2V}\right)}$$

and that the frequency of the sustained oscillation, in radians per second, is $\omega_c = \pi V/(2b + d)$.

Comment briefly on the advantages and disadvantages of this system compared with an alternative arrangement in which the hopper valve operates to control the rate of feed according to the equation

$$m(t) = K'\{W_1 - W(t)\}$$

where K' is a constant. (*Brunel Univ.*)

Chapter 7

The Bode Plot and Nichols Chart

The *Bode plot* is a plot of magnitude and angle of a transfer function against frequency. If the axes have logarithmic scales the plot can be approximated by straight-line segments, and this is of great practical value. The transfer function to be plotted will in general have the form

$$G(j\omega) = \frac{K\left[1 + \frac{2\zeta}{\omega_n} j\omega + \frac{1}{\omega_n^{2}}(j\omega)^2\right]^w}{(j\omega)^t(1 + j\omega\tau_1)^u(1 + j\omega\tau_2)^v}$$

where t, u, v, w can be positive or negative integers.

The magnitude on a logarithmic scale (decibels) is

$$\begin{aligned} 20\log_{10}|G(j\omega)| = 20\log_{10}\mathrm{K} - t20\log_{10}\omega - u20\log_{10}|1 + j\omega\tau_1| \\ -v20\log_{10}|1 + j\omega\tau_2| \\ + w20\log_{10}\left|1 + \frac{2\zeta}{\omega_n} j\omega + \frac{1}{\omega_n^{2}}(j\omega)^2\right| \end{aligned}$$

This shows that a complete magnitude plot can be obtained by adding plots of each factor.

Similarly the argument is

$$\begin{aligned} \underline{/G(j\omega)} = -t\underline{/90^\circ} - u\underline{/\tan^{-1}\omega\tau_1} - v\underline{/\tan^{-1}\omega\tau_2} \\ + w\underline{/\tan^{-1}\frac{2\zeta\omega/\omega_n}{1 + (\omega/\omega_n)^2}} \end{aligned}$$

and this can be obtained by adding the phase angles of the individual factors.

The basic terms that can appear are:

A constant, K

$(j\omega)^{\pm 1}$

$(1 + j\omega\tau)^{\pm 1}$

$[1 + 2\zeta(j\omega)/(\omega_n + (j\omega)^2/\omega_n^{2}]^{\pm 1}$

These factors lead to the following plots.

1. *The constant K*

This gives a constant magnitude, $20 \log_{10} K$, and zero phase shift.

2. $(j\omega)^{\pm 1}$

$$\text{Magnitude} = \pm 20 \log_{10} \omega$$

This is a straight line with a slope of 20 dB/decade (6 dB/octave) and having the value of 0 dB at $\omega = 1$ rad/s. The slope is positive if the term is in the numerator and negative if in the denominator. The phase shift is constant at $\pm 90°$.

3. $(1 + j\omega\tau)^{\pm 1}$

$$\text{Magnitude} = \pm 20 \log_{10} |1 + j\omega\tau|$$

This can be approximated by the asymptotes

$20 \log_{10} 1$, i.e. 0 dB, for $\omega\tau \ll 1$

$\pm 20 \log_{10} \omega\tau$ for $\omega\tau \gg 1$ (this has a slope of $\pm$ 20 dB/decade)

The asymptotes meet at the point where

$$20 \log_{10} \omega\tau = 0$$

i.e. $\omega = 1/\tau$, which is called the *corner frequency*, ω_c, or *break point*. The maximum error due to this approximation is 3 dB and occurs at the corner frequency.

The associated phase angle is $\pm \tan^{-1} \omega\tau$ The phase angle plot can be approximated by three straight-line segments:

For $\omega < 1/10\tau$, a phase shift of 0°.
For $\omega > 10/\tau$, a phase shift of $\pm 90°$.
For $1/10\tau < \omega < 10/\tau$, a straight line, slope $\pm 45°$ per decade passing through a phase angle of $\pm 45°$ at the corner frequency.

Examples of this type of plot of gain and phase angle are shown in Fig 7.1. for the transfer functions

$$G(j\omega) = 1 + 0{\cdot}1 j\omega$$

and

$$G(j\omega) = \frac{1}{1 + j\omega}$$

4. $[1 + 2\zeta(j\omega)/\omega_n + (j\omega)^2/\omega_n{}^2]^{\pm 1}$

If $\zeta > 1$ this can be reduced to two first-order factors. If $\zeta < 1$,

$$\begin{aligned}\text{Magnitude} &= \pm 20 \log_{10} |1 - (\omega/\omega_n)^2 + j2\zeta\omega/\omega_n| \\ &= \pm 20 \log_{10} [\{1 - (\omega/\omega_n)^2\}^2 + 4\zeta^2(\omega/\omega_n)^2]^{1/2}\end{aligned}$$

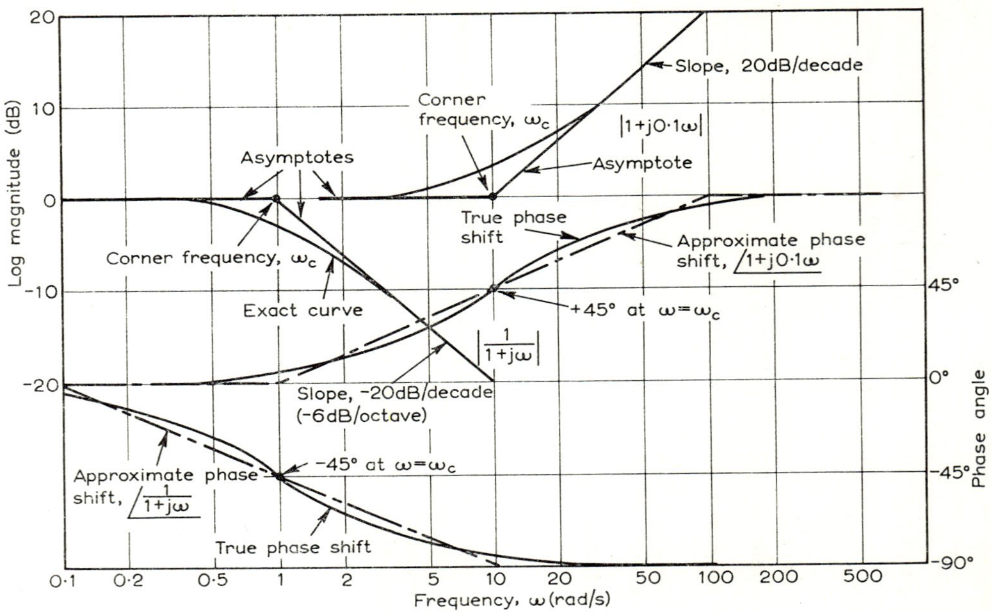

20
10
0
-10
-20
Log magnitude (dB)
Slope, 20dB/decade
Corner
frequency, ω_c
$|1+j0{\cdot}1\omega|$
Asymptotes
Asymptote
True phase
shift
Approximate phase
shift, $\angle 1+j0{\cdot}1\omega$
Corner frequency, ω_c
Exact curve
+45° at $\omega=\omega_c$
$\left|\frac{1}{1+j\omega}\right|$
Slope, −20dB/decade
(−6dB/octave)
−45° at $\omega=\omega_c$
Approximate phase
shift, $\angle \frac{1}{1+j\omega}$
True phase shift
45°
0°
−45°
−90°
Phase angle
0·1
0·2
0·5
1
2
5
10
20
50
100
200
500
Frequency, ω (rad/s)

FIG 7.1

This expression can be approximated by two straight-line segments with a break point at $\omega = \omega_n$. The segment below $\omega = \omega_n$ is constant at 0 dB while that above ω_n has a slope of ± 40 dB/decade.

The error due to these approximations depends upon the value of the damping ratio ζ and can be very large if ζ is small.

The phase shift is given by

$$\underline{/G(j\omega)} = \pm \tan^{-1} \frac{2\zeta\omega/\omega_n}{1 - (\omega/\omega_n)^2}$$

This can be approximated by three straight-line segments as follows:

For $\omega < \omega_n/10$, the phase shift is constant at 0°.
For $\omega > 10\omega_n$, the phase shift is constant at 180°.
For $\omega_n/10 < \omega < 10\omega_n$ the phase angle can be represented by a straight line, having a slope of $\pm 90°$/decade passing through the 90° point at $\omega = \omega_n$.

As with the magnitude plots, the error increases rapidly as $\zeta \to 0$.

Examples of this type of plot are shown in Fig. 7.2. Graphs (*a*) and (*b*) show the magnitude and phase shift of the function

$$1/[1 + j\omega + (j\omega)^2]$$

(a term which is also used in Problem 6.3). A simple calculation will show that this has a natural angular frequency of 1 rad/s and a damping ratio of 0·5. Graphs (*c*) and (*d*) show magnitude and phase for a quadratic function again having a natural frequency of 1 rad/s but with a damping ratio of 0·1.

The poles of the expression $1/(1 + s + s^2)$ are at $s = -0{\cdot}5 \pm j0{\cdot}86$. It is left as an exercise for the reader to show that the magnitude and phase response can be obtained from the pole-zero diagram by appropriate measurement along the imaginary axis, i.e. $s = j\omega$. An example of this procedure has already been given in Chapter 5.

Nichols Chart

This chart enables the closed-loop frequency response to be obtained by the use of the open-loop magnitude and phase angle plots (Bode diagram). The Nichols chart is a plot on Cartesian co-ordinates, of magnitude in decibels against angle in degrees. Contours of closed-loop gain and phase angle are superimposed on such a diagram, and correspond to M, α, loci of the polar plots, the origin corresponding to $(-1, j0)$. A typical Nichols chart showing the M, α curves will be found in the pocket inside the back cover.

The use of the Nichols chart and the Bode diagram can be illustrated by repeating the problem of the voltage regulator previously

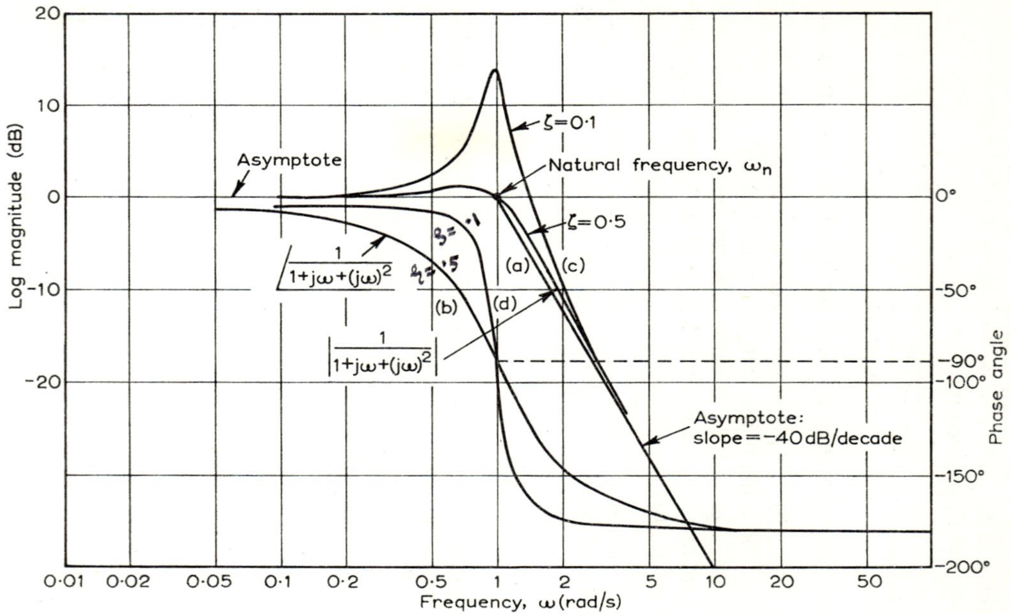
20
10
0
-10
-20
Log magnitude (dB)
0°
-50°
-90°
-100°
-150°
-200°
Phase angle
0·01
0·02
0·05
0·1
0·2
0·5
1
2
5
10
20
50
Frequency, ω (rad/s)
Asymptote
ζ=0·1
Natural frequency, ω_n
ζ=0·5
(a)
(b)
(c)
(d)
$\angle \frac{1}{1+j\omega+(j\omega)^2}$
$\left| \frac{1}{1+j\omega+(j\omega)^2} \right|$
Asymptote:
slope = −40 dB/decade

FIG 7.2

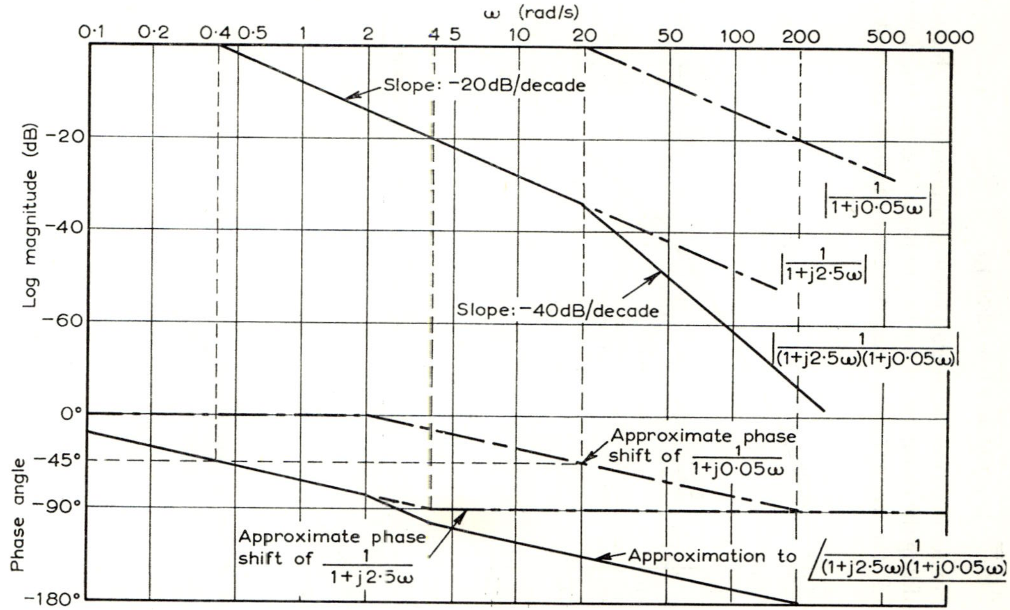
ω (rad/s)
0·1
0·2
0·4
0·5
1
2
4
5
10
20
50
100
200
500
1000
Log magnitude (dB)
-20
-40
-60
Slope: −20dB/decade
Slope: −40dB/decade
|1/(1+j0·05ω)|
|1/(1+j2·5ω)|
|1/((1+j2·5ω)(1+j0·05ω))|
Phase angle
0°
-45°
-90°
-180°
Approximate phase shift of 1/(1+j0·05ω)
Approximate phase shift of 1/(1+j2·5ω)
Approximation to ∠1/((1+j2·5ω)(1+j0·05ω))

FIG 7.3

analysed. The problem is to find the maximum value of $|(C/R)(j\omega)|$ for a loop gain of 50, and the frequency ω_m at which it occurs.

The open-loop transfer function is

$$\frac{C}{E}(j\omega) = \frac{50}{(1 + j0{\cdot}05\omega)(1 + j2{\cdot}5\omega)}$$

In this transfer function the overall log magnitude can be obtained by simply adding the ordinates of the two individual elements. Referring to Fig. 7.3, the asymptotes in each case are the 0 dB axis and a line

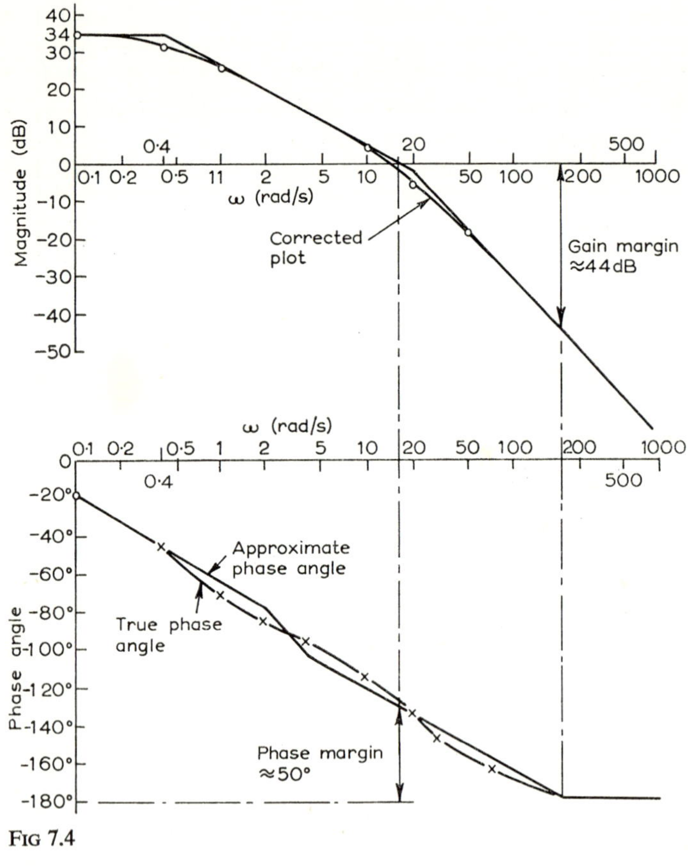

FIG 7.4

having a slope of -20 dB/decade. These meet at the corner frequency $\omega_c = 1/0{\cdot}05 = 20$ rad/s for the function $1/(1 + j0{\cdot}05)$. The other corner frequency is 0·4 rad/s. The combined straight line plot represents $20 \log_{10} |1/(1 + j0{\cdot}05\omega)(1 + j2{\cdot}5\omega)|$; the complete plots of magnitude and phase of $50/(1+j0{\cdot}05\omega)(1+j2{\cdot}5\omega)$ are shown in Fig. 7.4. To take account of the constant term, 50, the magnitude plot of Fig. 7.3 is increased by $20\log_{10}50 = 34$ dB. In practice this could easily be achieved by redrawing the origin at -34 dB in Fig. 7.3.

The straight-line approximation to the overall phase shift is shown in Fig. 7.3 and repeated on Fig. 7.4, together with the true phase angle. The latter is obtained from the table on page 179, with the magnitudes converted to decibels.

A definition of gain and phase margins has been given in Chapter 6. These can easily be obtained from the Bode plots, the gain margin being expressed in decibels in this case. For the system considered here, the gain margin is 44 dB and the phase margin approximately 50°.

The magnitudes and phase angles obtained from Fig. 7.4 have been plotted on a Nichols chart in Fig. 7.5. Examination of the M and α contours reveals a value of M_m of 1·2 dB and $\alpha = -54°$ at 14 rad/s. These compare favourably with the values obtained in Chapter 6.

Example 7.1

A unity-feedback control system has the configuration shown in Fig. 7.6 where U represents an extraneous signal appearing at the input to the plant. It is desired that the steady-state error resulting from an extraneous unit step signal $u(t) = 1$ shall be 0·1 units. Determine the value of controller gain K to meet this specification.

Construct the Bode diagram for the controller and plant, and comment on the stability and transient behaviour of the complete system. (*Leeds Univ.*)

Solution

In order to investigate the effect of the disturbance U on the error E the block diagram can be redrawn as in Fig. 7.7. From this diagram it can be seen that

$$\frac{E(s)}{U(s)} = \frac{-10}{(1+s)(1+0{\cdot}01s)+10K(1+0{\cdot}1s)}$$

The input U is a unit step, hence $U(s) = 1/s$, and

$$E(s) = \frac{-10}{s[(1+s)(1+0{\cdot}01s)+10K(1+0{\cdot}1s)]}$$

Application of the final-value theorem gives

$$\lim_{t\to\infty} e(t) = \lim_{s\to 0} s\,E(s) = -10/(1 + 10K)$$

The steady-state error must not be greater than 0·1 unit, giving $K = 9{\cdot}9$.

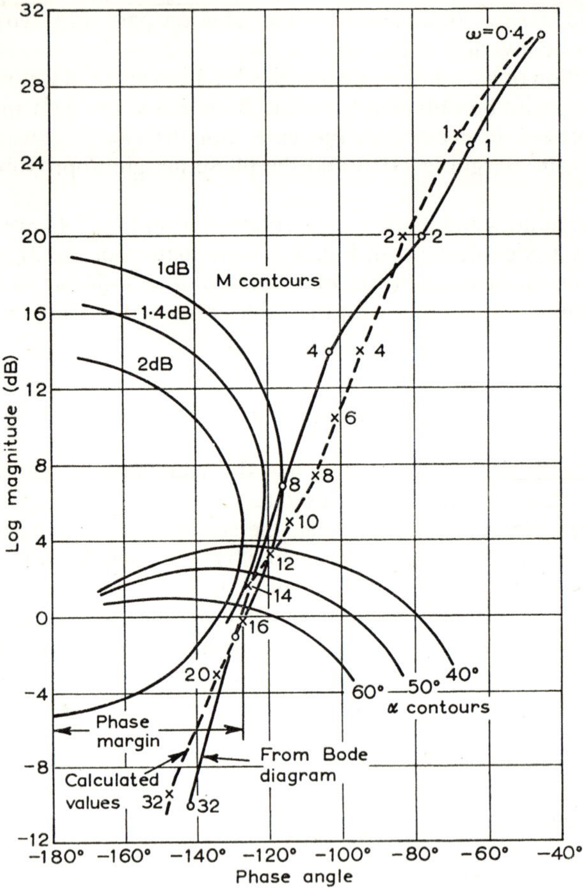

FIG 7.5

The Bode diagram is constructed for the open-loop frequency response function,

$$\frac{Q_0}{E}(j\omega) = \frac{99(1 + j0{\cdot}1\omega)}{(1 + j\omega)(1 + j0{\cdot}01\omega)}$$

and can be approximated by the sum of four factors:

1. The constant 99. This represents a gain of 39·5 dB and raises the complete plot by this amount.
2. The factor $(1 + j0{\cdot}1\omega)$. This can be approximated by two straight-line segments meeting at $\omega = 10$ rad/s. The first segment is constant at 0 dB, the second rises at 6 dB/octave.
3. The factor $1/(1 + j\omega)$. This again consists of two straight-line segments, the first constant at 0 dB, the second falling at 6 dB/octave. The break point is at $\omega = 1$ rad/s.
4. The factor $1/(1 + j0{\cdot}01\omega)$ is similar to the previous factor except that the break point is at 100 rad/s.

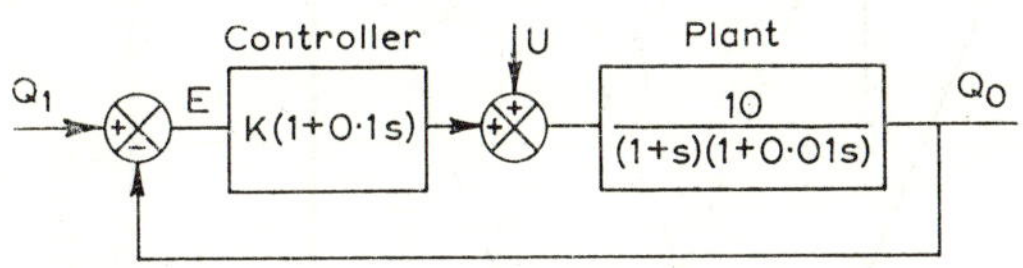

FIG 7.6

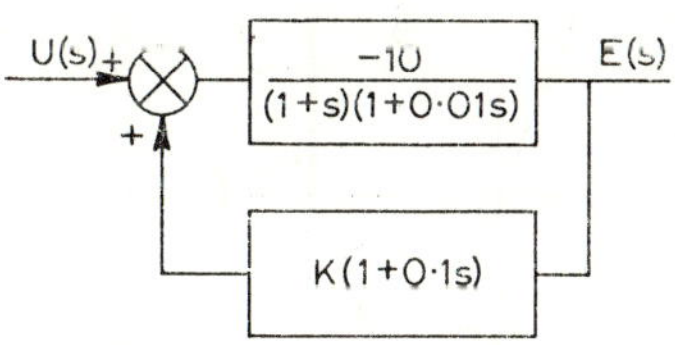

FIG 7.7

The first factor contributes no phase shift to the plot, while the remaining factors each give a phase shift that eventually reaches 90° (leading for factor 2 and lagging for factors 3 and 4). The approximate phase shift can be obtained by use of the method given in the introduction.

The approximate gain and phase plots are shown in Fig. 7.8. The system is very stable and approximates to a first order system at high frequencies. It will not have any overshoots in the transient behaviour.

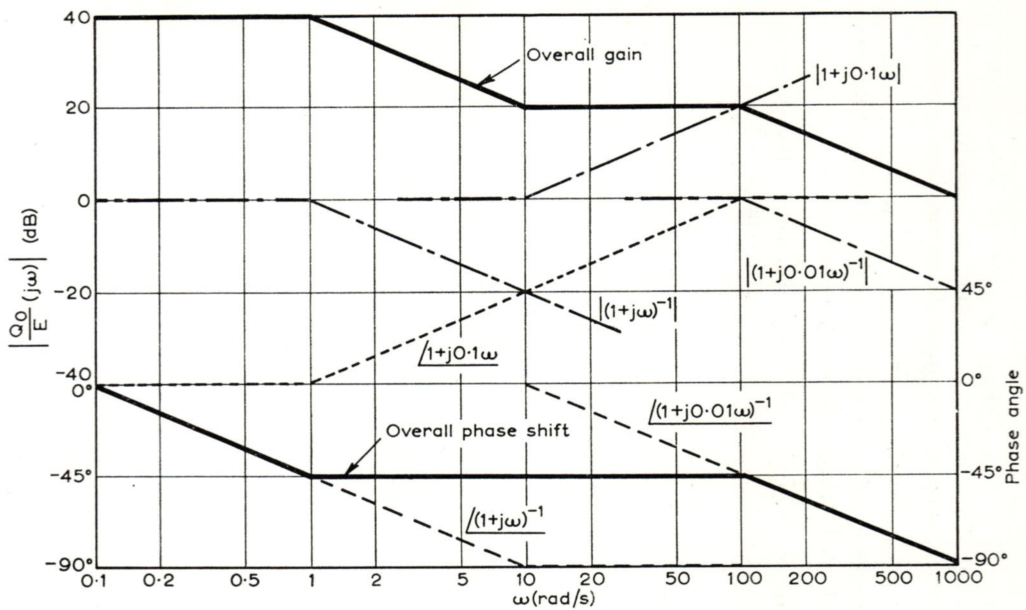
40
20
0
-20
-40
0°
-45°
-90°
$\left|\frac{Q_O}{E}(j\omega)\right|$ (dB)
Overall gain
$|1+j0{\cdot}1\omega|$
$|(1+j0{\cdot}01\omega)^{-1}|$
$|(1+j\omega)^{-1}|$
$\angle 1+j0{\cdot}1\omega$
$\angle(1+j0{\cdot}01\omega)^{-1}$
Overall phase shift
$\angle(1+j\omega)^{-1}$
45°
0°
-45°
-90°
Phase angle
0·1
0·2
0·5
1
2
5
10
20
50
100
200
500
1000
ω(rad/s)

FIG 7.8

Example 7.2

The control system in Fig. 7.9 has an open-loop attenuation diagram which may be approximated by straight lines as follows:

Frequency range	Slope of straight line
$0 < \omega < 20$ rad/s	-20 dB/decade
$20 < \omega < 100$ rad/s	-40 dB/decade
$100 < \omega < \infty$ rad/s	-60 dB/decade

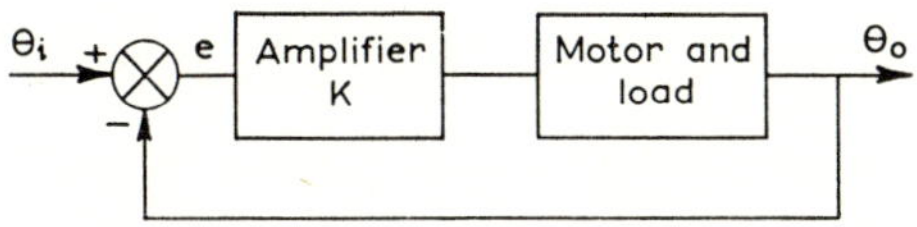

FIG 7.9

The line of slope -40 dB/decade crosses the 0 dB axis at $\omega = 50$ rad/s.

Write down the open-loop transfer function including values for the gain constant K and the appropriate time constants, and test for the stability of the closed-loop system.

If a series phase-advance network having the transfer function

$$\frac{(1 + 0{\cdot}1s)}{10(1 + 0{\cdot}01s)}$$

is introduced before the amplifier, estimate the new value of phase margin.

Comment on the effect of this compensator network on the steady-state error for a ramp input. (*Hatfield*)

Solution

The straight-line segments are plotted in Fig. 7.10. As the initial segment has a slope of 20 dB/decade this must indicate a term $1/j\omega$ in the transfer function. If this were the only term in the transfer function and the gain constant K was unity the segment would cross the 0 dB axis at $\omega = 1$ rad/s. Extension of this initial segment (as shown in Fig. 7.10) gives the crossing point at $\omega = 130$ rad/s. Hence

$$20 \log_{10} (K/130) = 0 \qquad \text{i.e.} \qquad K = 130$$

Two break points indicate two terms of the form $1/(1 + j\omega\tau)$ in the transfer function. The break points occur at $\omega = 20$ rad/s and $\omega = 100$ rad/s, giving values for τ of 0·05 and 0·01 s respectively. The complete transfer function is

$$\frac{130}{j\omega(1 + j0{\cdot}05\omega)(1 + j0{\cdot}01\omega)}$$

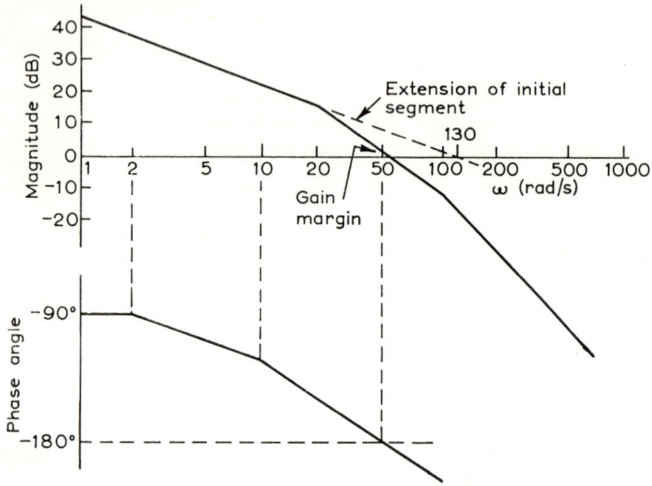

FIG 7.10

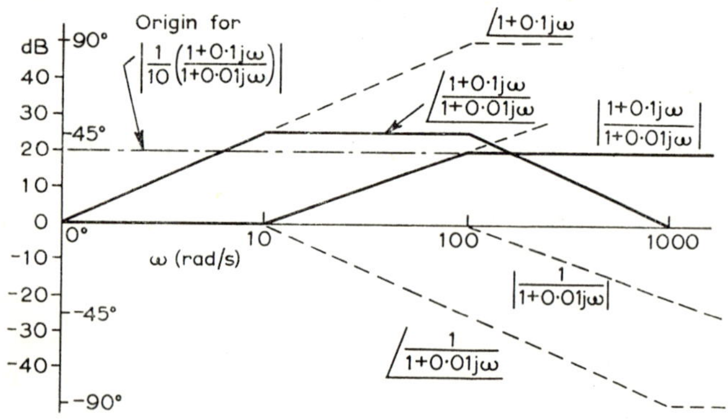

FIG 7.11

From the gain segments, the segments for approximate phase shift have been constructed. From these the gain margin is approximately −3 dB and the system is unstable. The Bode plot of the compensating network is shown in Fig. 7.11, and the modified system in Fig. 7.12. From this plot the phase margin is now 70°.

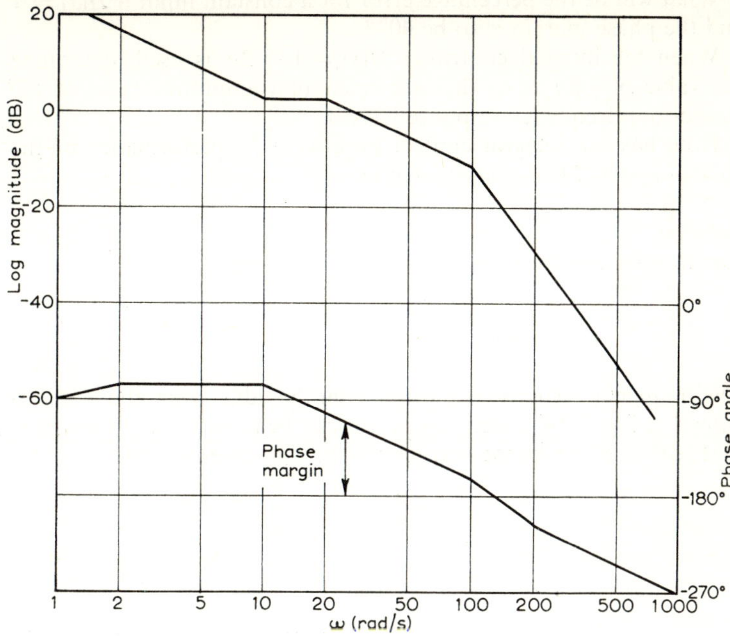

FIG 7.12

The compensator does not affect the type number of the system but does reduce the static loop gain. Hence the system will still follow a ramp input but the steady-state error is increased by a factor of 10.

Example 7.3

Fig. 7.13 represents a control system for which the controller characteristic K and $H(s)$ are to be chosen. Information about the transfer function $G(s)$ derived from frequency-response tests is given in the following table:

ω (rad/s)	0	0·5	1	2	4
$20 \log_{10} \lvert G(j\omega)\rvert$	0	−1	−4	−11	−22
$\underline{/G(j\omega)}$	0	−48°	−86°	−135°	−186°

Draw $G(j\omega)$ on the Nichols chart. What can be said about the transfer function $G(s)$?

What will be the percentage error for a constant input if $H(s) = 1$ and the phase margin is to be 40°?

When the integral controller $H(s) = 1 + 2/s$ is used, determine the value of gain K so that the maximum amplitude ratio of the closed-loop frequency response is 4 dB.

How has the integral control improved the performance of the control system? (*Edinburgh Univ.*)

Solution

The transfer function $G(s)$ must be of the form

$$G(s) = \frac{1}{(1 + s\tau_1)(1 + s\tau_2)(1 + s\tau_3)}$$

The frequency response is plotted from the table given above and is shown in Fig. 7.14. From this plot it can be seen that the gain must be 12 dB to give a phase margin of 40°. This gives $K = 4$.

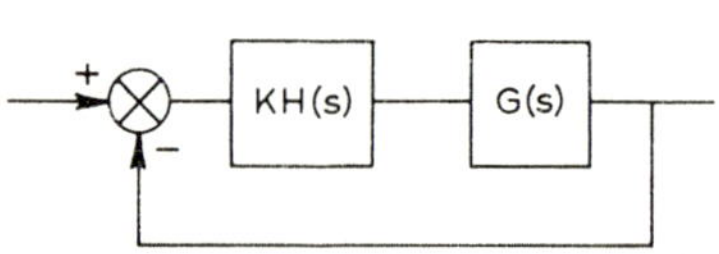

FIG 7.13

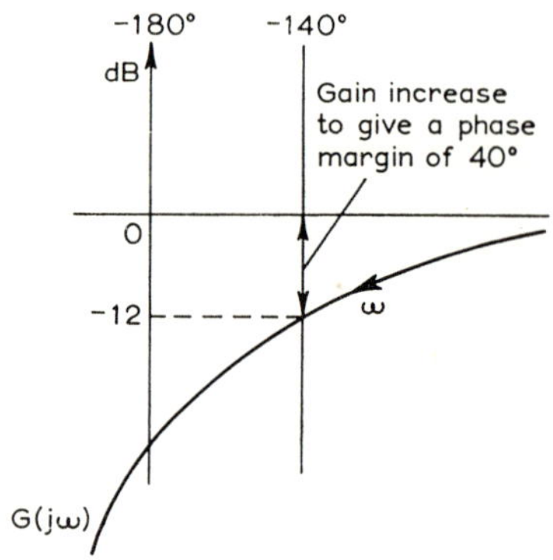

FIG 7.14

The steady-state error of this system to a constant input is

$$\frac{1}{1 + K} = 0{\cdot}2$$

so that the percentage error is 20%.

The integral controller has a transfer function given by

$$H(j\omega) = \frac{2 + j\omega}{j\omega} = \frac{2(1 + 0{\cdot}5j\omega)}{j\omega}$$

The gain and phase angle of this expression are given in the following table:

ω (rad/s)	0	0·5	1	2	4
$20 \log_{10} \lvert 1 + 0{\cdot}5j\omega \rvert$ $20 \log_{10} \lvert 1/j\omega \rvert$ $20 \log_{10} 2$	0 ∞ 6	0·26 6·0 6	1·0 0 6	3 −6 6	7 −12 6
$20 \log_{10} \left\lvert \frac{2(1 + 0{\cdot}5j\omega)}{j\omega} \right\rvert$	∞	12·2	7·0	3·0	1
$\underline{/(1 + 0{\cdot}5j\omega)}$ $\underline{/1/j\omega}$	0 −90°	+14° −90°	+26·5° −90°	+45° −90°	+63° −90°
$\underline{/(1 + 0{\cdot}5j\omega)/j\omega}$	−90°	−76°	−63·5°	−45°	−27°

Alternatively the response of the controller can be obtained from its Bode plot. This is shown in Fig. 7.15. The overall transfer function

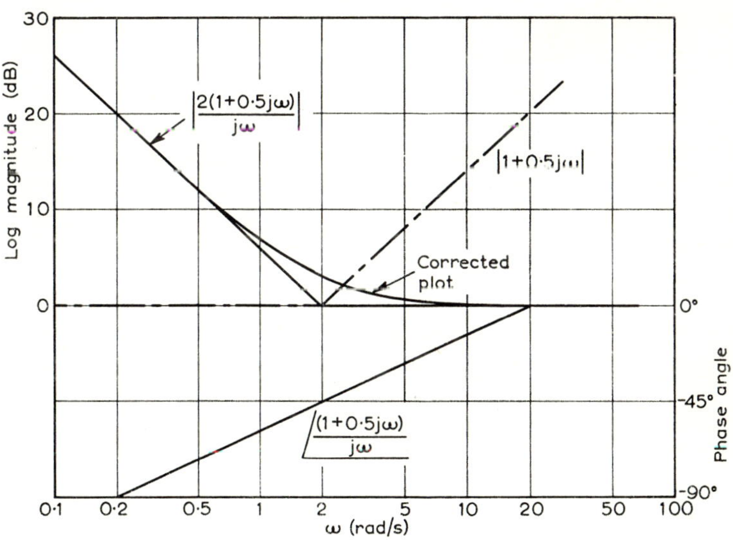

FIG 7.15

$G(j\omega)\,H(j\omega)$ is obtained by combining these results with those for $G(j\omega)$. The results are given in the following table and plotted in Fig. 7.16.

ω (rad/s)	0	0·5	1	2	4
$\lvert G(j\omega)H(j\omega)\rvert$	∞	11·2 dB	3·0 dB	−8·0 dB	−21 dB
$\underline{/G(j\omega)H(j\omega)}$	−90°	−124°	−149°	−180°	−213°

In order to obtain a maximum amplitude ratio of 4 dB under closed-loop conditions, the frequency response must be moved vertically until it is tangential to the 4 dB *M* contour. This occurs for a vertical shift of 4 dB, and corresponds to a value of *K* of 0·63.

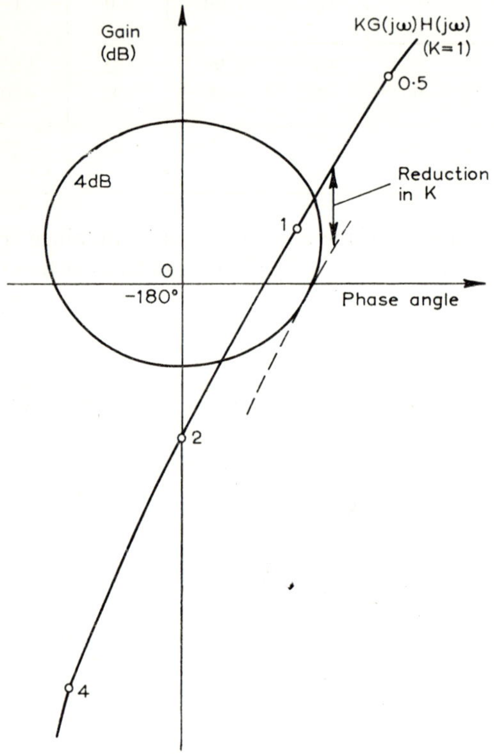

FIG 7.16

The integral control has converted the system to type 1 and it will have zero error for constant input signals.

Example 7.4

A stabilized voltage supply of rated output 12 V, 0·5 A is shown in block form in Fig. 7.17.

The input voltage, V_1, is from a transformer, bridge rectifier and reservoir capacitor and is sufficiently above V_3 to let the controlling transistors, A, operate normally. V_1 comes from an effective source impedance of $16\,\Omega$, it has 0·4 V r.m.s. of 100 Hz ripple, and it is directly proportional to variations in the alternating input supply voltage.

The controlling transistors, A, are two cascaded emitter followers where $i_3 = +4\,000 i_2$. High-frequency effects in the transistors reduce this gain by 20 dB/decade for frequencies above 16 kHz, and by

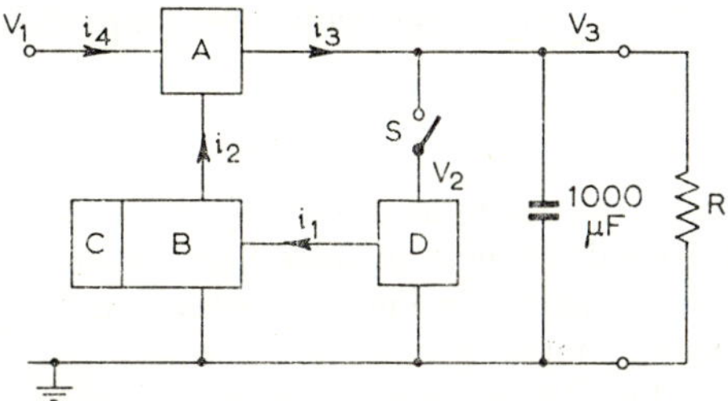

FIG 7.17

40 dB/decade for frequencies over 160 kHz. These cause the appropriate phase lag. The internal impedance in the line i_4 to i_3 is $1{\cdot}2\,\Omega$. The differential amplifier, B, has an output/input relation of $i_2 = -100 i_1$.

The Zener reference, C, has a small temperature coefficient such that the differential amplifier output current, i_2, has a temperature coefficient of $-0{\cdot}05$ mA/°C.

The potential divider, D, has a transfer admittance of 250 μS between the input at V_2 and the output i_1; i.e. a small change of input δV_2 volts, will give a change of output of $\delta V_2/4\,000$ *A*.

The output voltage, V_3, is developed across a 1 000 μF capacitor and a load resistor, R, that takes rated current.

Where time-constants and temperature coefficients are not mentioned for any components in the circuit, these are assumed to be negligible. The impedances of paths other than those mentioned can be assumed to be high. Static voltage levels in the circuit are set by the ratio of the potentiometer [voltage divider], D, and the voltage of the Zener reference, C; they need not otherwise be considered.

With the switch S open, find the change in output voltage, V_3, caused by a small change in V_2 in the form of a transfer function. This is the open-loop transfer function; draw this as a Bode diagram to determine if the circuit would be stable for the load shown with S closed, and if so, by what gain and phase margins. Use the asymptotic approximation for the Bode diagram and the following phase angle data for a circuit having a single time-constant:

Frequency, ω, rad/s	$1/10\tau$	$1/3{\cdot}16\tau$	$1/\tau$	$3{\cdot}16/\tau$	$10/\tau$
Phase angle	6°	18°	45°	72°	84°

On open loop, what will be the d.c. output impedance, the r.m.s. ripple output, the change in output voltage for a 10% change in the a.c. supply, and the change in output voltage for a 20°C rise in ambient temperature? What are these with the loop closed? (*Sheffield Univ.*)

Solution

In Fig. 7.17, V_1, V_2 and V_3 are taken as direct voltage levels while i_1, i_2, i_3, and i_4 represent small changes in operating currents. The block diagram of the system is shown in Fig. 7.18, where the changes in V_1, V_2 and V_3 have been denoted by v_1, v_2 and v_3 respectively. All the small signals shown in Fig. 7.18 have been written as functions of s to denote that they are Laplace transforms.

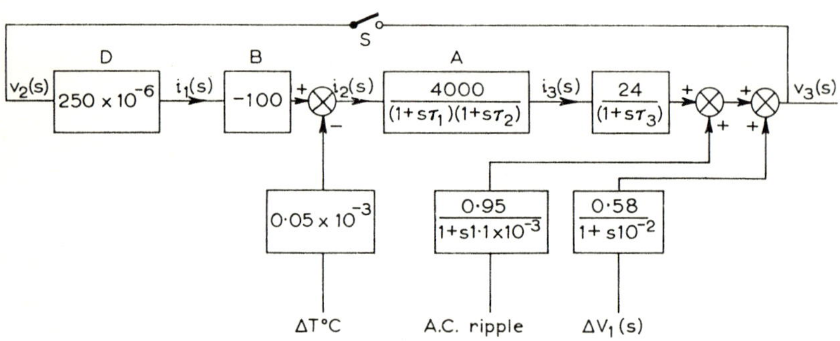

FIG 7.18
$\tau_1 = 10^{-5}$ s $\quad \tau_2 = 10^{-6}$ s $\quad \tau_3 = 24 \times 10^{-3}$ s

The relationship between $v_3(s)$ and $i_3(s)$ is obtained from the fact that $v_3(s) = i_3(s)Z(s)$ where $Z(s)$ is the operational impedance of the load R and the 1 000 μF capacitor in parallel. As the load takes rated current its value must be 24 Ω.

$$v_3(s) = \frac{24/1\,000 \times 10^{-6} s}{24 + 1/1\,000 \times 10^{-6} s}\, i_3(s)$$

$$= \frac{24}{1 + 24 \times 10^{-3} s}\, i_3(s)$$

From the block diagram the relationship between $v_3(s)$ and $v_2(s)$ can be written (ignoring disturbances)

$$\frac{v_3(s)}{v_2(s)} = \frac{-2\,400}{(1 + 10^{-5}s)(1 + 10^{-6}s)(1 + 24 \times 10^{-3}s)}$$

The Bode diagram of this expression is plotted in Fig. 7.19 (the phase angle has been obtained with the aid of the data given in the question). Measurements on this plot give a gain margin of 20 dB and a phase margin of 40°.

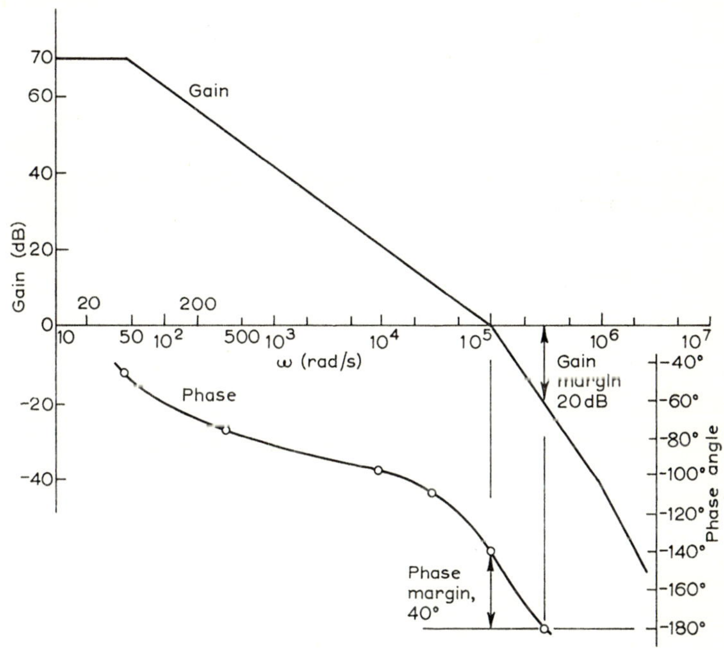

FIG 7.19

The effect of disturbances due to a.c. ripple and supply voltage changes can be obtained using the circuits shown in Fig. 7.20. These give the transfer functions shown in block diagram form in Fig. 7.18. The effect of these disturbances on the output when S is open can be obtained as follows.

A 20°C rise in temperature will give $i_2(t)$ as a fall of 1 mA, giving $i_3(t)$ as a fall of 4 A; hence a 96 V fall as $v_3(t)$. Thus the open-loop condition is impractical.

400 mV ripple at 100 Hz will give the magnitude of the change of V_3 as

$$v_3 = \frac{0{\cdot}95 \times 400}{\sqrt{[1 + (2\pi 100 \times 1{\cdot}1 \times 10^{-3})^2]}} = 312\,\text{mV}$$

A 10% change in supply voltage will cause a 10% change in output voltage, which is 1·2 V.

By inspection the output impedance is 17·2 Ω.

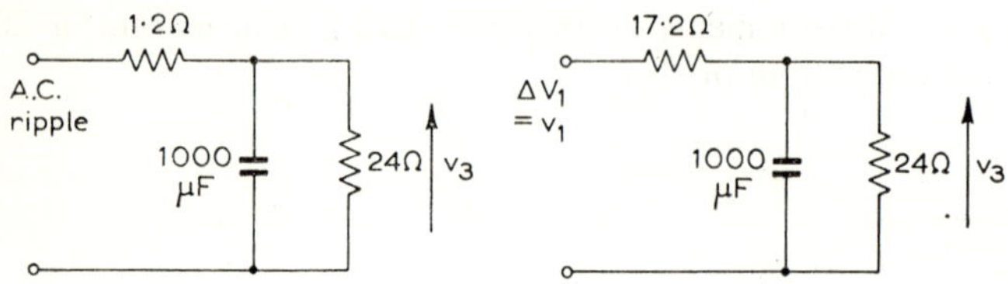

FIG 7.20

With the switch S closed the effects of these disturbances are multiplied by a factor

$$\frac{1}{1 + \text{open-loop gain}}$$

(see Chapter 3). The output impedance is

$$\frac{17{\cdot}2}{1 + 2400} = 7{\cdot}2\,\text{m}\Omega$$

The temperature change causes a change in V_3 given by

$$\frac{96}{1 + 2400} = 40\,\text{mV}$$

The change in V_3 due to the ripple is now given by

$$\frac{312}{1 + \text{open-loop gain at } f = 100\,\text{Hz}}$$

The open-loop gain at $\omega = 100\,\text{Hz}$ can be obtained directly from the plot of Fig. 7.19:

$$\left|\frac{v_3}{v_2}(j\omega)\right| = 160$$

The change in V_3 under the closed-loop condition is therefore 2 mV.
The change in V_3 due to a 10% change in the a.c. supply is given by

$$\frac{1{\cdot}2}{1 + 2400} = 0{\cdot}5\,\text{mV}$$

Example 7.5

Fig. 7.21 indicates the transfer functions of the elements of a hydraulic control system. Construct the Bode diagram of the system and hence

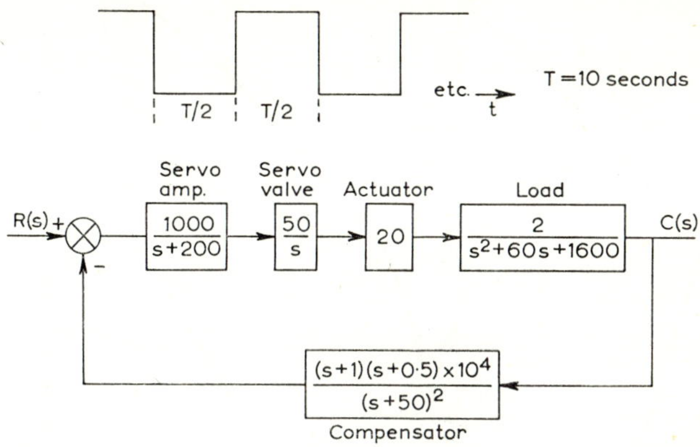

FIG 7.21

discuss the performance in relation to a repetitive input signal of the form shown in Fig. 7.21. (*Birmingham Univ.*)

Solution

In view of the complexity of the system it would be extremely laborious to obtain a Bode plot by any of the methods previously used. An approximate method is as follows.

The transfer function of a system having a forward-path transfer function $G(s)$ and feedback-path transfer function $H(s)$ is

$$\frac{C}{R}(s) = \frac{G(s)}{1 + G(s)\,H(s)}$$

When $|G(s)\,H(s)| \ll 1$, then $\dfrac{C}{R}(s) \approx G(s)$

When $|G(s)\,H(s)| \gg 1$, then $\dfrac{C}{R}(s) \approx \dfrac{1}{H(s)}$

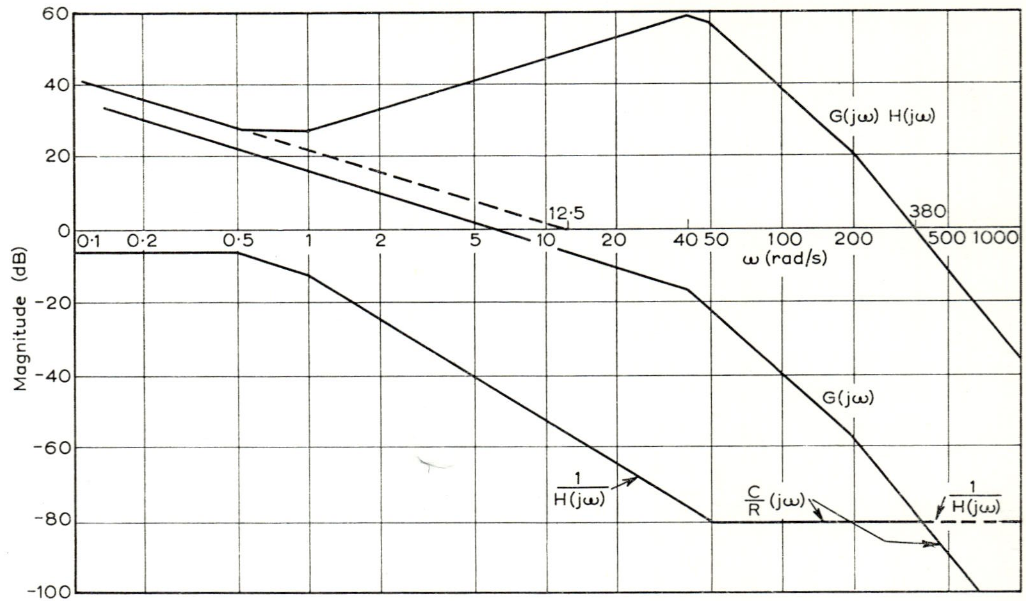

60
40
20
0
-20
-40
-60
-80
-100
Magnitude (dB)
0·1
0·2
0·5
1
2
5
10
20
40 50
100
200
500 1000
ω (rad/s)
12·5
380
G(jω) H(jω)
G(jω)
1/H(jω)
C/R (jω)
1/H(jω)

FIG 7.22

In this example,

$$G(s)\,H(s) = \frac{2 \times 10^6}{s(s + 200)(s^2 + 60s + 1\,600)} \; \frac{10^4(s + 1)(s + 0{\cdot}5)}{(s + 50)^2}$$

$$= \frac{12{\cdot}5(1 + s)(1 + 2s)}{s(1 + 0{\cdot}005s)(1 + 0{\cdot}02s)^2(1 + 60s/1\,600 + s^2/1\,600)}$$

$$G(s) = \frac{6{\cdot}25}{s(1 + 0{\cdot}005s)(1 + 60s/1\,600 + s^2/1\,600)}$$

$$\frac{1}{H(s)} = \frac{0{\cdot}5(1 + 0{\cdot}02s)^2}{(1 + s)(1 + 2s)}$$

The magnitudes of these functions have been plotted, using straight-line approximations, in Fig. 7.22. The quadratic factor of $G(s)\,H(s)$ has $\omega_n = 40$ rad/s and $\zeta = 0{\cdot}75$, and has been represented as a single break point with a slope of 40 dB/decade following this point. From this plot $|GH| = 1$ at $\omega = 380$ rad/s and the $1/H(s)$ plot can be used up to this frequency to represent $(C/R)(j\omega)$.

The bandwidth of the system is 0·5 rad/s. The square-wave applied voltage can be considered as a Fourier series, consisting of a fundamental plus odd harmonics. The fundamental frequency is 0·1 Hz (0·62 rad/s); hence the harmonics will be attenuated.

An estimate of the rise time can be obtained if the system is approximated by a simple lag:

$$\frac{C}{R}(s) = \frac{1}{(1 + 2s)}$$

This will give an exponential rise to a step input with a time-constant of 2 s. The output will reach within 1 % of its final value in 10 s. The approximate shape of the output is shown in Fig. 7.23.

PROBLEMS

7.1 The following experimental results of gain and angular frequency were obtained for a control system under open-loop conditions:

Angular frequency (rad/s)	0·1	0·3	1	3	10	30	100	300
Gain (dB)	60·0	50·5	40·5	32	25·5	15·5	−2·0	−21·0

Assuming two break points, find the approximate open-loop transfer function. Establish whether or not the system will be

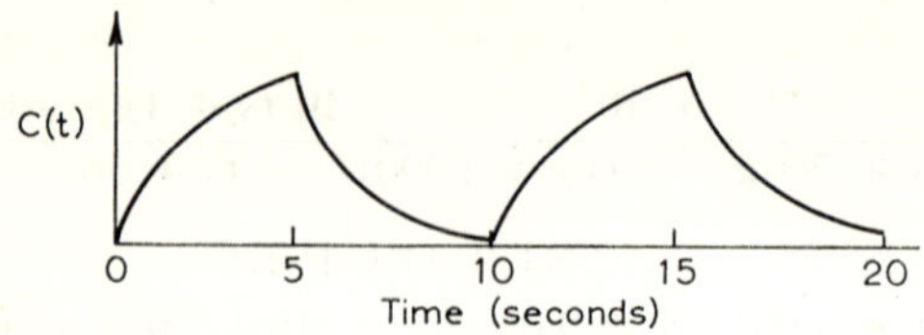

FIG 7.23

unstable when the loop is closed. (Use the chart provided.) (*Scottish Assoc.*)

Ans. $\dfrac{100(1 + 0{\cdot}2j\omega)}{j\omega(1 + 0{\cdot}05j\omega)^2}$

7.2 Show how an approximate plot of gain (dB) against logarithm of frequency of the function $(1 + j\omega\tau)^{\pm m}$ may be obtained.

The open-loop transfer function of a control system is

$$KG(j\omega) = \frac{K_A}{j\omega(1 + 0{\cdot}5j\omega)(1 + 0{\cdot}1j\omega)}$$

Plot the approximate Bode-diagram gain-frequency characteristic.

Find the phase lag at the two break points and assuming a linear characteristic between these two points, explain how the plot indicates that this system is unstable.

If K_A is to be changed to make the gain margin 10 dB, find the required value of K_A and the corresponding phase margin. (*Scottish Assoc.*)

Ans. At break point 1, phase shift = 148°; at break point 2, phase shift = 211°; K_A = 11 db, 23°.

7.3 The schematic diagram of Fig. 7.24 shows a system for controlling the speed of a d.c. motor. Draw a block diagram of the system, inserting all the relevant transfer functions given that the characteristics of the components are:

Amplifier gain K (not specified)
Generator (*a*) gain, 2·5 V/V
(*b*) field-circuit time-constant, 1·0 second
(*c*) armature resistance, 0·01 Ω
(*d*) armature inductance to be neglected
Motor (*a*) rating: 450 V, 78·5 rad/s
(*b*) armature resistance, 0·02 Ω
(*c*) armature inductance to be neglected
(*d*) total inertia at motor shaft, 200 kg-m^2.
Tachogenerator constant, 20 V per 105 rad/s.

Assuming an open-loop gain of 8, determine the gain and phase margins of the system by the use of logarithmic plots relating magnitude and phase angle with angular frequency (rad/s). (Straight-line approximations may be used.) (*Lanchester*)*

Ans. 37 dB, 48°.

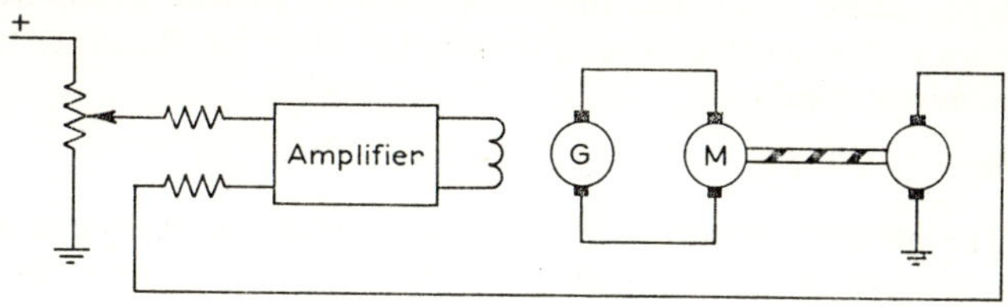

FIG 7.24

7.4 The following transfer function refers to a series stabilizing network for a closed-loop control system

$$G(s) = \frac{(1 + s\tau_2)(1 + s\tau_3)}{(1 + s\tau_1)(1 + s\tau_4)}$$

$\tau_1 = 2\,\text{s}$, $\tau_2 = 0{\cdot}5\,\text{s}$, $\tau_3 = 0{\cdot}25\,\text{s}$, $\tau_4 = 0{\cdot}0625\,\text{s}$.

Draw an approximate Bode attenuation diagram for the network, and from it deduce the frequency at which this network introduces no additional phase-shift into the system.

Assuming that at the above frequency the output impedance of the network is to be of the order of 0·5 MΩ, determine a set of values for the elements and draw the network circuit diagram. Discuss the advantages of a network of this type when used for stabilizing a closed-loop system. (*Rugby*)

Ans 2·8 rad/s

7.5 The results of an open-loop frequency response test using a transfer function analyser are given in the table. Find the open-loop transfer function of the system being tested

Frequency (rad/s)	0·1	0·3	1	3	10	30	100	300	1000	3000	10000
Log gain (dB)	41·6	32·1	21·7	13	7·3	5·8	3·0	−4·1	−15	−28·6	−48·1

(*Hatfield*)

Ans $$\frac{12(1 + j\omega 0{\cdot}18)}{j\omega(1 + j\omega 0{\cdot}01)(1 + j\omega 0{\cdot}0005)}$$

7.6 Explain how the open-loop frequency response in the form of a log-gain/log-frequency diagram can be used to assess the stability of a closed-loop control system. Using this technique estimate the value of τ such that a system with an open-loop transfer function

$$\frac{\theta_o}{\theta_i - \theta_o} = \frac{16(1 + s/2)}{s^2(1 + s\tau)^2}$$

shall be stable when the feedback loop is closed. Compare this estimate with the value obtained using the Hurwitz–Routh criterion. (*Manchester Univ.*)

Ans. $< 0{\cdot}125$ s.

7.7 The network shown in Fig. 7.25 is used in the series compensation of certain servomechanisms. Derive the transfer function

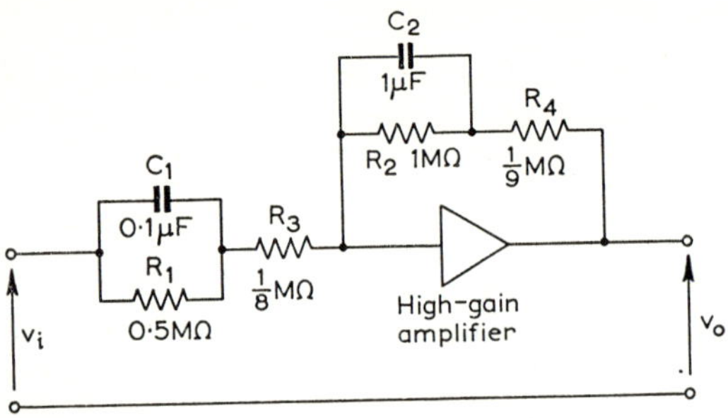

FIG 7.25

$V_o(s)/V_i(s)$ and plot its approximate straight-line attenuation diagram and its phase locus for the component values given. Comment on the shape of the loci, and discuss the advantages of this form of compensation. What advantages are obtained by the use of an active network? (*Hatfield*)

Ans. $$\frac{1{\cdot}78(1 + j\omega 0{\cdot}1)(1 + j\omega 0{\cdot}05)}{(1 + j\omega)(1 + j\omega 0{\cdot}01)}$$

7.8 Discuss the advantages of using Bode diagrams for control-system design based on frequency characteristics. A control system with unity negative feedback has a closed-loop transfer function

$$\frac{\theta_o}{\theta_i} = \frac{K}{s(s + 1)(s + 5) + K}$$

Determine, using Bode plots and Routh's criterion, the limit on the value of K for maintaining stability. Determine also the gain K required to stabilize the system with a phase margin of 30°, and the corresponding value of the velocity-error constant K_V.

The uncompensated open-loop phase characteristic is given by

ω (rad/s)	0·2	0·5	1·0	1·5	2·0	5·0	10·0
$\underline{/G(j\omega)}$	−103·5°	−122°	−146°	−163°	−175°	−213·5°	−237·5°

Indicate, on the Bode plots, the effects on the log-magnitude characteristic of adding a lag-lead compensating network in the forward path of the system with $K = 100$. (*Cambridge Univ.*)

Ans. 30, 8, 1·6.

7.9 Show how approximate frequency-response characteristics, for gain and phase shift, can be constructed for the transfer function $(1 + \tau s)$ and $1/(1 + s\tau)$ by the Bode method.

Determine the errors at the break points. Construct the Bode diagram for a control system defined by

$$KG(s) = \frac{K}{s(1 + s)(1 + 0{\cdot}1s)}, \quad H(s) = 1$$

Use the diagram to determine the range of K for stable operations, and also the phase margin for $K = 10$. (*Liverpool Univ.*)

Ans. $K < 11$; for $K = 10$ phase margin $\approx 1°$.

7.10 Discuss, with reference to the stability of linear systems, the methods of Hurwitz–Routh, Nyquist and Bode. The open-loop transfer function of a fourth-order position control system is given by

$$H(s) = \frac{\omega_0^2(1 + \tau s)}{s^2(1 + \alpha\tau s)^2}$$

If $\tau\omega_0 = 1$, determine, by any one of the above methods, the maximum value of α for which the system remains stable. (*I.E.E.*)

Ans. 0·4.

7.11 (*a*) Explain the conditions for the limit of stability of a unity feedback system by the use of (i) the Nyquist criterion and diagram, (ii) the Bode plot. Show briefly how these criteria are related to the closed loop transfer function

$$\frac{\theta_o}{\theta_i}(s) = \frac{KG(s)}{1 + KG(s)}$$

(*b*) Show how the use of M-circles on the Nyquist diagram allows the degree of stability of such systems to be assessed.

(*c*) A particular system has the following open-loop frequency response:

ω (rad/s)	17	30	40	55	75	140	240
$KG(j\omega)$ (dB)	+30	+20	+15	+10	+5	−6	−14
Phase lag, ϕ	−150°	−165°	−172°	−175°	−178°	−185°	−188°

By plotting on the Nichols chart provided, ascertain the degree of stability. (*Brunel Univ.*)

Ans. −2 dB, −2°.

7.12 It was required that an analysis be made of a feedback position control system arranged as shown in Fig. 7.26. The error-detecting

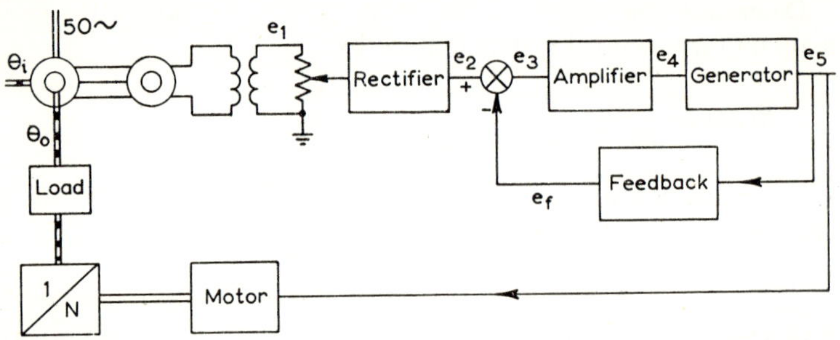

FIG 7.26

elements were 50 Hz synchros whose output was 4 V/rad error. This was stepped up by 4:1 by a transformer and controlled by the potentiometer [voltage divider] K. The rectifier and smoothing stage had a static gain of 4, the frequency-dependent function being that of a simple lag with a break point at $\omega = 50$ rad/s. The amplifier and generator were part of a minor feedback loop, the feedback function being given by

$$\frac{e_f}{e_5}(s) = \frac{4}{s + 20}$$

The amplifier had a gain of 10 and the generator transfer function was

$$\frac{e_5}{e_4}(s) = \frac{4(s + 20)}{(s + 40)}$$

The high-torque motor drove the load via a gearbox such that the load speed was given by

$$s\theta_o = 10e_5$$

Obtain the open-loop transfer function in factored form, and by sketching the asymptotic Bode plot on log paper, estimate the gain margin if $K = 1$. Assume that the phase shift has reached its final value away from the break-point frequency. (*Brunel Univ.*)

Ans. $\dfrac{2\,560K(1 + 0{\cdot}05s)}{s(1 + 0{\cdot}02s)(1 + 0{\cdot}005s)}$; 10 dB

7.13 A negative-feedback amplifier has a closed-loop frequency response function of

$$\frac{10}{1 + 0{\cdot}01j\omega}$$

If the feedback factor H is 0·099, determine

(i) The forward open-loop frequency response function.
(ii) The forward open-loop gain at zero frequency.
(iii) At what frequencies the open-loop and closed-loop amplitudes fall by 3 dB from that at zero frequency.

Comment briefly on the results obtained in (iii).

(iv) If the zero-frequency gain of the open-loop amplifier is reduced by 50%, find the new closed-loop frequency response function and comment briefly on the significance of the result. (*Bangor*)

Ans. (i) $\dfrac{1\,000}{1 + j\omega}$; (ii) 1 000; (iii) 0·16 Hz, 16 Hz; (iv) $\dfrac{9{\cdot}9}{1 + 0{\cdot}02j\omega}$

7.14 A system has the forward-path transfer function $G(s) = 10/s(1 + 2s)$ and the feedback transfer function $H(s) = 1$. Using straight-line approximations, draw the Bode diagram showing the magnitude and the phase characteristic as a function of frequency. Use the Nichols chart to obtain the closed-loop frequency response.

What is the resonant frequency of the system? Can this system be made unstable by altering the gain constant? (*Lanchester*)

Ans. 2·25 rad/s.

7.15 Explain how the open-loop frequency response, in the form of a log-gain/log-frequency diagram, can be used to assess the stability of a closed-loop control system. Using this technique, estimate the maximum value of τ_1 such that a system with an open-loop transfer function

$$\frac{\theta_o}{\theta_i - \theta_o} = \frac{1 + s\tau}{s\tau(1 + s\tau_1)^3}$$

shall be stable when the feedback loop is closed. Compare this estimate with the value obtained using the Hurwitz–Routh criterion, and explain any difference observed between the two results. (*Queen's Univ., Belfast*)

Ans. $\tau_1 = 1{\cdot}55\tau$.

7.16 Determine the transfer function of the passive network shown in Fig. 7.27 (*a*). Using a logarithmic scale, draw graphs of the

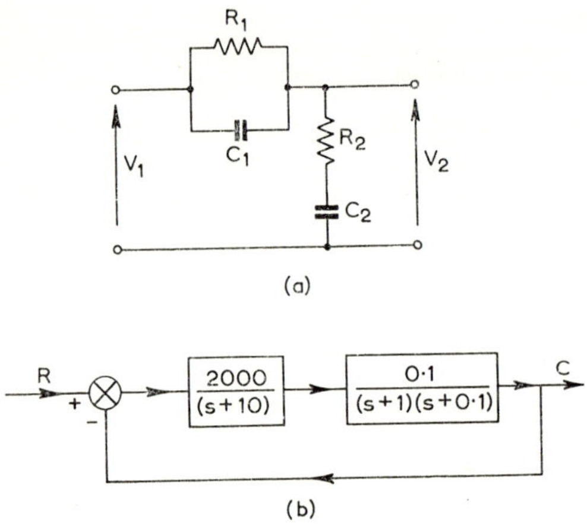

FIG 7.27

$R_1C_1 = 0{\cdot}2\,\text{s} \quad R_2C_2 = 10\,\text{s} \quad R_2C_1 = \frac{1}{44{\cdot}91}\,\text{s}$

approximate gain and phase shift of the network against frequency.

The speed control system shown in Fig. 7.27 (*b*) requires full open-loop gain if the accuracy specification is to be met. Predict the effect on the system stability of inserting the network of Fig. 7.27 (*a*) in series with the amplifier (assume that the network is not loaded in any way). (*Southampton Univ.*)

Ans. $\dfrac{V_2}{V_1} = \dfrac{(1 + 0{\cdot}2s)(1 + 10s)}{(1 + 0{\cdot}02s)(1 + 100s)}$; 28 dB, 53°.

7.17 For the case of a closed-loop unity-feedback system, define the term "*M*-circles." Indicate how they are derived and explain the use the M_{max} criterion. A particular plant has the following open-loop frequency response:

ω	2	4	5	6	8	12	16
M	5	3·2	1·8	1·26	0·8	0·64	0·32
M (dB)	14	10·1	5·1	2	−2	−3·9	−10
ϕ	−10°	−20°	−30°	−50°	−90°	−105°	−130°

If the above plant is incorporated in a unity-feedback control system with the controller simply a gain constant of unity, ascertain the phase margin for the system, using the Nichols chart.

If now the controller comprises a perfect integrator of unity gain in cascade with a variable gain amplifier, determine the value of the amplifier gain constant if the new system is to have an M_{max} of 6 dB. What are now the phase margin and gain margin? (*Brunel Univ.*)

Ans. 110°; $K = 5$; 36°, 6 dB.

7.18 An industrial process has the forward-path transfer function:

$$G(s) = \frac{10}{1 + s + s^2}$$

and the feedback-path transfer function

$$H(s) = \frac{1}{1 + 2s}$$

Draw an approximate log-magnitude versus log-frequency diagram for the closed-loop system. State clearly all approximations made to obtain this plot. How must the above transfer functions be arranged so that a Nichols chart could be used to obtain a more accurate closed-loop plot? (*Lanchester*)

Ans. Plot exhibits peak of 10dB at $\omega = 1{\cdot}7$ rad/s.

7.19 A control system has an open-loop transfer-function of the form

$$G(s)\,H(s) = \frac{K}{(1 + s)(1 + 0{\cdot}2s)(1 + 0{\cdot}1s)}$$

In an attempt to improve the performance of the system simple phase-lead and phase-lag sections are added in turn, the resultant transfer functions then being:

$$\text{Phase-lead, } G(s)\,H(s) = \frac{6(1 + 0{\cdot}077s)}{(1 + s)(1 + 0{\cdot}2s)(1 + 0{\cdot}1s)(1 + 0{\cdot}0077s)}$$

$$\text{Phase-lag, } G(s)\,H(s) = \frac{40(1 + 2s)}{(1 + s)(1 + 0{\cdot}2s)(1 + 0{\cdot}1s)(1 + 20s)}$$

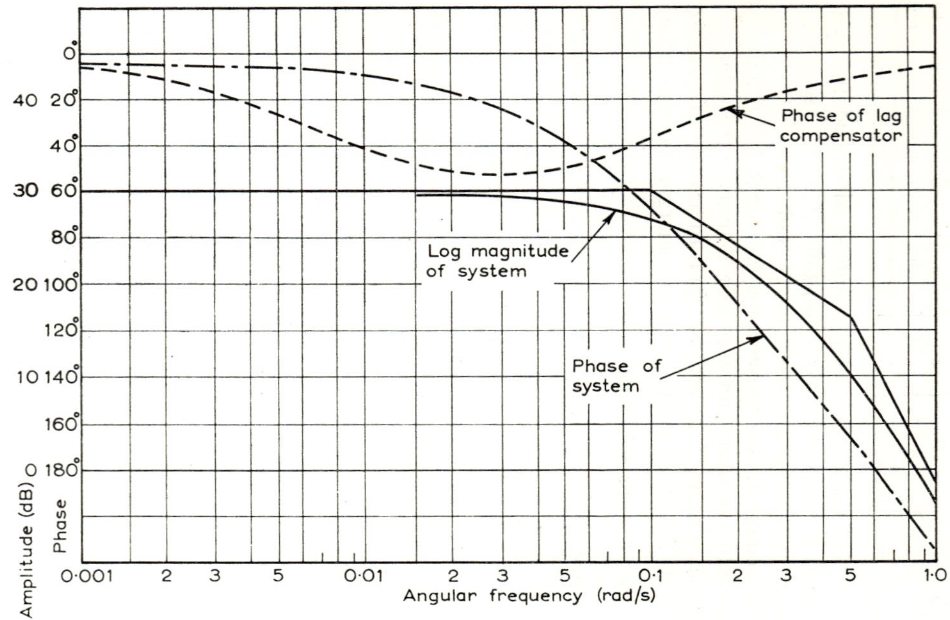

0°
40 20°
40°
30 60°
80°
20 100°
120°
10 140°
160°
0 180°
Amplitude (dB)
Phase
0·001
2
3
5
0·01
2
3
5
0·1
2
3
5
1·0
Angular frequency (rad/s)
Phase of lag compensator
Log magnitude of system
Phase of system

FIG. 7.28

Values of $|G(j\omega)H(j\omega)|$ and argument $G(j\omega)H(j\omega)$ are given in the table below.

Lead

ω (rad/s)	1·0	1·5	2	3	5	7	10	15	20	30
$\lvert G(j\omega)H(j\omega)\rvert$	3·6	2·7	2·2	1·5	0·75	0·42	0·22	0·09	0·07	0·02
Arg $G(j\omega)H(j\omega)$	−56°	−73°	−85°	−105°	−130°	−144°	−157°	−169°	−177°	−187°

Lag

ω (rad/s)	1·0	1·5	2	3	5	7	10	15	20	30
$\lvert G(j\omega)H(j\omega)\rvert$	3·2	2·5	1·9	1·2	0·6	0·33	0·15	0·06	0·02	0·01
Arg $G(j\omega)H(j\omega)$	−89°	−95°	−110°	−128°	−157°	−176°	−195°	−215°	−228°	−241°

Check that the phase and gain margins derived from the tabulated results are correct (allowing for errors associated with the use of uncorrected Bode gain plots).

If the closed-loop system operates with unity-gain feedback, i.e. $H(s) = 1$, derive values for the closed-loop gains corresponding to the values of ω in the table, and plot gain against log ω.

Comment on the static accuracy and the dynamic response to step-function inputs, expected for the two systems. Compare also the responses of the two to a periodic disturbance (*a*) originating within the loop, (*b*) superimposed on the input signal, if the frequency of the disturbance is (i) less than 0·08 Hz, (ii) about 0·25 Hz. Assuming that the spacing of the corner frequencies of each network is reasonable, comment on the particular values chosen.

If the original transfer function had been of the form

$$G(s)H(s) = \frac{K}{(1 + 0{\cdot}02s)(1 + 0{\cdot}2s)(1 + s)}$$

state which type of network would have been more suitable and explain why. (*Sheffield Univ.*)

Ans. Lead network, 26 dB, 54°; lag network, 11 dB, 40°.

Closed-loop gain:
Lead network, $\omega = 1$, −1·4 dB; $\omega = 4$, 0 dB.
Lag network, $\omega = 1$, −0·6 dB; $\omega = 4$, 3 dB.

Lag compensator has greater static accuracy due to higher gain constant.

(*a*) (i) Lag network preferable, (ii) little difference.
(*b*) (i) Little difference, (ii) lead network preferable.

7.20 Derive the transfer function of an electric lag compensator comprised of two resistors R_1 and R_2 and a capacitor C, expressing it in terms of a time constant τ and a factor α.

Fig. 7.28 shows the Bode plot for the open loop of a system with three poles. Assuming unity feedback write down the forward-path transfer function. State whether the closed-loop system is stable, and determine the phase margin.

The performance is to be improved by adding a lag compensator

$$G_c(j\omega) = \frac{1 + j\omega 10}{1 + j\omega 100}$$

The additional phase lag introduced by this lag network is shown on the Bode plot. Plot on Fig. 7.28 the log-magnitude and phase curves (including straight-line approximations) for the compensated system. Discuss the improvement in performance. (*Nottingham Univ.*)

Ans. Unstable, phase margin $-25°$; stable, gain margin 10 dB, phase margin 30°.

7.21 Discuss the use of Bode logarithmic plots in the design of closed-loop control systems and comment on the limitations of steady-state frequency response design techniques.

A control system has an open-loop frequency response $KG(j\omega)$ of the form

$$KG(j\omega) = \frac{K}{j\omega(1 + 0{\cdot}1j\omega)(1 + 0{\cdot}2j\omega)(1 + j\omega)}$$

with

ω (rad/s)	0·6	1	2	3	4	10
$\underline{/G(j\omega)}$	−131°	−152°	−187°	−209°	−227°	−283°

Using a Bode logarithmic plot, with corrections, find:

(*a*) The value of the steady-state error coefficient which makes the system marginally stable on closed-loop.

(*b*) The value of the steady-state error coefficient which results in a gain margin of 10 dB, and also the corresponding phase margin.

(*c*) The steady-state output of the closed-loop system, set for a margin of 10 dB, when its input is $5 \sin (t - \pi/6)$.

(*Cambridge Univ.*)

Ans. (*a*) 3·2; (*b*) 1, 38°; (*c*) $7 \sin (t - 2{\cdot}3)$.

Appendix

Complex Number Table

This table gives the modulus, M, and argument, θ, of the complex number $1 + jx$.

$$Me^{j\theta} = 1 + jx, \text{ where } M = (1 + x)^{1/2} \text{ and } \theta = \tan^{-1}(x).$$

x	M	$\theta°$	x	M	$\theta°$	x	M	$\theta°$
0·05	1	3	1·8	2·06	61	5·2	5·3	79
0·10	1·00	6	1·85	2·12	62	5·4	5·49	79
0·15	1·01	8	1·90	2·15	62	5·6	5·69	80
0·20	1·02	11	1·95	2·18	63	5·8	5·89	80
0·25	1·03	14	2·00	2·23	63	6·0	6·08	81
0·30	1·04	17	2·1	2·32	64	6·2	6·28	81
0·35	1·06	19	2·2	2·42	66	6·4	6·48	81
0·40	1·08	22	2·3	2·51	66	6·6	6·68	81
0·45	1·09	24	2·4	2·60	67	6·8	6·87	82
0·50	1·12	27	2·5	2·69	68	7·0	7·07	82
0·55	1·14	29	2·6	2·79	69	7·2	7·27	82
0·60	1·17	31	2·7	2·88	70	7·4	7·47	82
0·65	1·19	33	2·8	2·97	70	7·6	7·67	82
0·70	1·22	35	2·9	3·07	71	7·8	7·87	83
0·75	1·25	37	3·0	3·16	72	8·0	8·06	83
0·80	1·28	39	3·1	3·26	72	8·2	8·26	83
0·85	1·31	40	3·2	3·35	73	8·4	8·46	83
0·9	1·34	42	3·3	3·45	73	8·6	8·66	83
0·95	1·38	43	3·4	3·54	74	8·8	8·86	83
1·00	1·41	45	3·5	3·64	74	9·0	9·05	84
1·05	1·45	46	3·6	3·74	75	9·2	9·25	84
1·1	1·49	48	3·7	3·83	75	9·4	9·45	84
1·15	1·52	49	3·8	3·94	75	9·6	9·65	84
1·2	1·55	50	3·9	4·02	76	10	10·05	84
1·25	1·62	51	4·0	4·12	76	12	12·0	85
1·3	1·64	52	4·1	4·22	76	15	15·0	86
1·35	1·67	53	4·2	4·32	77	20	20·0	87
1·40	1·72	54	4·3	4·42	77	30	30	88
1·45	1·76	55	4·4	4·51	77	40	40	89
1·50	1·8	56	4·5	4·61	77	50	50	89
1·55	1·88	57	4·6	4·71	78	60	60	89
1·60	1·89	58	4·7	4·81	78	70	70	89
1·65	1·93	59	4·8	4·90	78	80	80	89
1·7	1·97	59	4·9	5·00	78	90	90	89
1·75	2·01	60	5·0	5·1	79	100	100	89
						∞	∞	90

Bibliography

It would be beyond the scope of a book such as this to give an extensive coverage of the published textbooks on control engineering. We give here a short list of some books which we think the student may find useful as an introduction to the subject. The mature student will not require assistance from us as he will no doubt have his own favourites; we apologize if we omit such texts from the list.

1. SENSICLE, A., *Introduction to Control Theory for Engineers* (Blackie, 1968). A good introduction to the subject, with a useful chapter on analogue computing, recommended for H.N.C. students.
2. STOCKDALE, L. A., *Servomechanisms* (Pitman, 1962). A useful introductory textbook. The root-locus approach is not covered, but there are good chapters on frequency response.
3. TAYLOR, P. L., *Servomechanisms* (Longmans, 1960). An introductory textbook for electrical engineering students.
4. WARD, J. R., and STRUM, R. D., *Signal Flow Graph in Linear System Analysis* (Prentice-Hall). Gives an introduction to the signal flow graph and the use of the Mason gain formula.
5. WEBB, C. R., *Automatic Control* (McGraw-Hill, 1964). Written for mechanical engineers in the main, but gives a good introduction to the pole-zero approach.
6. NIXON, F. E., *Principles of Automatic Controls* (Prentice Hall, 1957). A very useful book; although published in 1957 it gives a clear account of frequency response and root-locus techniques.
7. CHESTNUT, H., and MAYER, R. W., *Servomechanisms and Regulating System Design*, Vol. 1, 2nd ed. (Wiley, 1963). Recommended for the degree student; gives a sound introduction to the subject.
8. THALER, G. J., and BROWN, R. G., *Analysis and Design of Feedback Control Systems*, 2nd ed. (McGraw-Hill, 1960). Covers sampled-data and non-linear systems in addition to an introduction to the design of linear systems.
9. D'AZZO, J. J., and HOUPIS, C. H., *Feedback Control System Analysis and Synthesis*, 2nd ed. (McGraw-Hill, 1960). One of the best books available for what might be called "conventional"

control techniques. Also includes chapters on non-linear systems, a.c. systems and analogue computation.

10. Gibson, J. E., and Tuteur, F. B., *Control System Components* (McGraw-Hill, 1958). Deals with a wide range of component characteristics.
11. Shinners, S. M., *Control System Design* (Wiley, 1964). Will be useful to the degree student since it covers modern control engineering topics as well as the more conventional control design techniques.

Index